6th EDITION

Save Time ▸ Save Money ▸ Take Anywhere

Updated to the Current Codes Including the IRC, UPC, and UMC

Code Check®

Plumbing & Mechanical

Drainage
Traps & Vents
Water Supply
Gas Piping
Water Heaters
Fixtures
Ventilation
Ducts
Appliances
Gas Appliance Vents
Fireplaces
Furnaces

An Illustrated Guide to the Plumbing and Mechanical Codes

INTERNATIONAL CODE COUNCIL®

By Douglas Hansen, Skip Walker & Redwood Kardon
Illustrated by Paddy Morrissey, Kaia Mathewson & Douglas Hansen

Code✓Check Plumbing & Mechanical® Sixth Edition

By DOUGLAS HANSEN, SKIP WALKER & REDWOOD KARDON
Illustrations by Paddy Morrissey, Kaia Mathewson & Douglas Hansen

 ISBN 978-1-64155-208-0

Updated to the 2021 International Residential Code, Uniform Plumbing Code & Uniform Mechanical Code

Code Check Plumbing & Mechanical 6th edition is an illustrated reference guide to code requirements and common violations in residential plumbing and mechanical systems. The main codes referenced in this book are the *2021 International Residential Code*, published by the International Code Council, the *2021 Uniform Plumbing Code*, and the *2021 Uniform Mechanical Code*. For most topics, these codes are in agreement. These are the most widely used codes throughout the United States. NFPA 54, the *National Fuel Gas Code*, is the basis for the fuel gas provisions of the IRC, UPC, and UMC. Other referenced codes used in the book are listed in Table 1 (**T1**) below.

Model codes are updated on a 3-year cycle. In most areas, the 2021 code cycle will remain in effect for 3 to 6 years after the cover date. Significant changes from the previous code editions are highlighted in the text so that this book can be used in areas still using older code editions. Minor changes and those that only affected numbering (not substance) are not highlighted.

Energy codes vary greatly from one area to another and may modify or overrule the requirements shown in this book. Before beginning any project, check with your local building department to determine the codes and editions that apply in your area. Some jurisdictions modify the model codes. The codes also reference standards, many of which are maintained by the organizations in Table 2 (**T2**).

TABLE 1 — CODES USED IN THIS BOOK

Organization	Edition	Code
ASHRAE	2019	ASHRAE 62.2 Ventilation and Acceptable Indoor Air Quality in Residential Buildings
ICC	2021	International Residential Code (IRC)
ICC	2021	IPSDC—International Private Sewage Disposal Code
IAPMO	2021	Uniform Mechanical Code (UMC)
IAPMO	2021	Uniform Plumbing Code (UPC)
NFPA	2020	NFPA 31 Standard for the Installation of Oil-Burning Equipment
NFPA	2021	NFPA 54 National Fuel Gas Code
NFPA	2020	NFPA 58 Liquefied Petroleum Gas Code
NFPA	2020	NFPA 70 National Electrical Code
NFPA	2019	NFPA 211 Standard for Chimneys, Fireplaces, Vents, and Solid Fuel–Burning Appliances

TABLE 2 — STANDARDS ORGANIZATIONS

Acronym	Name
ACCA	Air Conditioning Contractors of America
ANSI	American National Standards Institute
ASHRAE	American Society of Heating, Refrigerating & Air-Conditioning Engineers
ASME	American Society of Mechanical Engineers
ASSE	American Society of Sanitary Engineering
ASTM	ASTM International (formerly American Society for Testing & Materials)
CSA	CSA Group (Canadian Standards Association)
IAPMO	International Association of Plumbing & Mechanical Officials
ICC	International Code Council
NFPA	National Fire Protection Association
NSF	National Sanitation Foundation
SMACNA	Sheet Metal & Air Conditioning Contractors' National Association
UL	UL (formerly Underwriters Laboratories)

KEY TO USING THIS BOOK

Large amounts of code information are condensed here by using several "shorthand" conventions. Many terms are abbreviated, as shown on the following page.

Each rule described in Code Check begins with a checkbox and ends with code citations. Where there are two columns of citations, the first one is from the IRC and the second one from the UPC or UMC, as noted at the top of the columns of code references. See this example from ***p. 4***:

Inspections	**21 IRC**	**21 UPC**
☐ All piping below slab tested before casting concrete___	109.1.2	105.1

This section is saying that piping must be tested before being covered by concrete. The IRC code reference is 109.1.2 & the UPC reference 105.1.

References to figures and tables are preceded by an **F** or a **T** as in the following examples from ***p. 13***:

☐ Trap seal min. 2 in., max. 4 in. **F21** ________________	3201.2	1005.1
☐ Size trap for fixture per **T8** ________________________	3201.7	1003.3

Figure 21 illustrates an example of this code rule, and the size is per Table 8.

Code changes are highlighted by showing the code citation in a different color, and by the superscript character that references a further explanation at the bottom of the page, as in this example from ***p. 9:***

☐ Max. 5 WCs on 3-in. horizontal or vertical drain **T10** ______	n/a	T703.2[2]

The UPC changed the maximum allowed number of WC's on a 3 in. drain, as explained in the following note that appears at the bottom of that page:

2. Previous UPC edition limited 3-in. pipe to 4 WCs on vertical drain & 3 on horizontal drain.

An "X" inside a code reference stands for the word "exception," as in this example from ***p. 15:***

☐ Increase 1 pipe size if >1/3 of vent is horizontal___________	n/a	904.2X

This line is saying that this rule is not applicable ("n/a") in the IRC, and in the UPC it is section 904.2 Exception.

When a text line ends with the letters EXC, it means that an exception follows in the next line, as in this example from ***p. 13***:

☐ Fixture tailpiece max. 24-in. vertical distance EXC **F20** __	3201.6	1001.2
• CW standpipes 18–42 in. (UPC: 18–30 in.) **F74** ___	2706.1.2	804.1

This line says that the maximum height of a fixture tailpiece (the vertical distance between the fixture outlet and its trap) is 24 inches in both codes, with an exception for a clothes washer standpipe. Notice also that the maximum height of the standpipe is not the same in the two codes, and the UPC size is shown in parentheses.

The information in this book is provided for informational purposes only and is not a substitute for the full text of the referenced codes. It should not be considered to be a substitute for the enforceable interpretation of the local building department.

***Benjamin Franklin** was chosen as the main character in our illustrations for several reasons. Franklin's insatiable curiosity, scientific genius, and civic-mindedness drove him to study fire safety, safe exiting, public sanitation, improved heating methods, and, of course, electricity.*

In 1752, he brought the first bathtub to America. After designing a more comfortable model, he took it with him on his travels to Europe.

CODE CHECK: *"YOUR KEY TO THE CODES."*

For updates, additional information on the codes, seminars, and online resources, visit: **www.codecheck.com**

ABBREVIATIONS

1&2FD = 1- & 2-family dwellings
AAV = air admittance valve
ABS = acrylonitrile-butadiene-styrene drain (black plastic pipe)
AC = air conditioning
ACH = air changes per hour
AFF = above finished floor
AHJ = Authority Having Jurisdiction
AL = aluminum
AMI = in accordance with MFR's instructions
AMM = alternative materials, design, & methods
AVB = atmospheric vacuum breaker
BO = building official
BT = bathtub
Btu = British thermal unit(s)
C = centigrade
Cat. = Category (appliance vent category: Cat. I, Cat. II, Cat. III, or Cat. IV)
cfm = cubic feet per minute
CI = cast iron
CO = cleanout
CPE = chlorinated polyethylene
CPSC = Consumer Product Safety Commission
CPVC = chlorinated PVC pipe
CSST = corrugated stainless-steel (gas) tubing
cu. = cubic, as in cu. ft.
Cu = copper
CW = clothes washer
CW&V = combination waste & vent
DFU = drainage fixture unit
DW = dishwasher
DWV = drain, waste & vent
e.g. = for example (exempli gratia)
EGC = equipment grounding conductor
EXC = exception to rule follows in the next line
F = Fahrenheit
FAU = forced-air unit
Fe = ferrous (iron or steel)
FLR = flood level rim
ft. = foot/feet
FVIR = flammable vapor ignition-resistant
ga = gauge
gal. = gallon(s)
GB = gypsum board
GFCI = ground-fault circuit interrupter
GPF = gallons per flush
gpm = gallons per minute
HDPE = high-density polyethylene
hr. = hour(s)
IBC = International Building Code
IFGC = International Fuel Gas Code
IMC = International Mechanical Code
in. = inch(es)
IPC = International Plumbing Code
k = 1,000 (1kBtu = 1,000Btu)
KS = kitchen sink
L&L = listed & labeled
lav = lavatory sink
lb. = pound(s)
LL = lot line
LP = liquefied petroleum (LP gas)
LT = laundry tray
max. = maximum
MFR = manufacturer
min. = minimum
MP = medium pressure
n/a = not applicable
NEC = National Electrical Code
NP = not permitted
NRTL = nationally recognized testing laboratory
o.c. = on center
O.D. = outside diameter
p. = pages, as in "***see p. 5***"
PB = polybutylene (water tubing)
PE = polyethylene (water or gas tubing)
PE-RT = polyethylene (raised temperature)
PEX = crossed-link polyethylene tubing
PP = polypropylene plastic tubing
PRR = pressure-reducing regulator
PRV = pressure-relief valve
psf = pounds per square foot
psi = pounds per square inch
psig = pounds per square inch gauge
PVC = polyvinyl chloride pipe
RP = reduced-pressure principle backflow preventer
SDC = Seismic Design Category
SFD = single-family dwelling
spec = specification
sq. = square, as in sq. ft.
SS = stainless steel
temp = temperature
TPRV = temperature & pressure-relief valve
w/ = with
w/o = without
WC = water closet (toilet)
WH = water heater
WSFU = water supply fixture unit
Zi = zinc, galvanized

TABLE OF CONTENTS

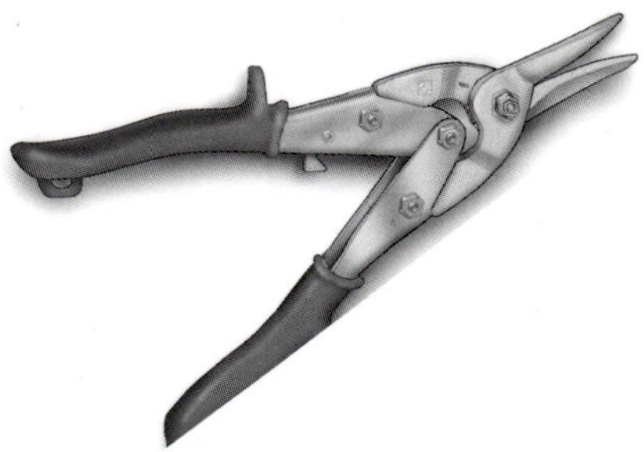

PLUMBING PERMITS & REQUIREMENTS

Materials or systems that do not conform to the codes (**T1**) or standards (**T2**) adopted by the local jurisdiction can be considered for approval on a case-by-case basis using the provisions for alternative materials, designs, and methods of construction (AMM). The permit applicant must provide documentation showing the proposed method to be equivalent to the intent of the code. If accepted by the BO, the approval would be limited to the property covered under the application; it does not mean the code is different than written for other properties in that jurisdiction.

Required Permits	21 IRC	21 UPC
☐ Permits required for installation, alteration, repair, replacement, or remodeling of a plumbing system EXC	105.1	104.1
• Stopping of leaks or clearing of stoppages	105.2	104.2
☐ If work so exempted results in necessity to replace or rearrange pipes, valves, or fixtures, a permit must be obtained	105.1	104.1
☐ Exempt work must still be done in accordance with code	105.2	103.1

Alternative Materials, Designs & Methods

☐ BO has authority to approve AMM if design satisfactory & meets intent of the code & is equivalent in terms of fire resistance, strength, effectiveness, durability, quality & safety	104.11	301.3
☐ Approved AMM is specific only to project in its scope	104.11	301.3
☐ When BO does not approve AMM, response must be written & reason stated for why it was not approved	104.11	n/a

Inspections

☐ No covering or concealing work prior to inspection	109.1.2	105.1
☐ All piping below slab tested before casting concrete	109.1.2	105.1
☐ Underground inspection before backfill	109.1.2	105.2
☐ Rough-in prior to covering or concealment	109.1.2	105.2
☐ Final inspection upon completion	109.1.6	105.2

Existing Buildings

☐ Existing installations that met code in effect at the time of construction allowed to remain if posing no hazard	102.7	102.2
☐ Additions, alterations & repairs to comply w/ rules for new construction w/o requiring existing areas to be brought to current code	102.7.1	102.4
☐ Not OK for additions, alterations & repairs to make existing building less compliant, unsafe, unsanitary, or overloaded	102.7.1	102.4
☐ Rehabilitation of existing pressure piping systems (epoxy lining) per ASTM F2831	n/a	320.1

MECHANICAL PERMITS & REQUIREMENTS

Electrical Requirements	21 IRC	20 NEC
☐ Receptacle within 25 ft. of appliance	3901.12	210.63
☐ Receptacle on same level as appliance	3901.12	210.63A
☐ Crawlspace furnace requires light w/switch at access	1305.1.3.3	210.70C
☐ Attic furnace requires light w/switch at access	1305.1.2.1	210.70C
☐ Individual circuit required for central heating	3703.1	422.12
☐ No other equipment on central heating circuit EXC	3703.1	422.12
• Associated pumps, humidifiers, air cleaners & AC	3703.1	422.12X1&2

Permits & Interpretations	21 IRC	21 UMC
☐ New installations, alterations & repairs require permits EXC	105.1	104.1
• Portable equipment & minor replacement parts	105.2	104.2
☐ BO may accept AMM that is equivalent in strength, quality, effectiveness, fire resistance, durability & safety	104.11	302.2
☐ Written response required for rejected AMM applications	104.11	local

Conflicts between Codes

☐ Installations not addressed by IRC to comply w/ IMC	1301.1	n/a
☐ If general & specific rules in conflict, specific rule prevails	102.1	102.1
☐ When codes in conflict, the more stringent prevails	102.1	102.1
☐ Plumbing code prevails if conflicting w/ mechanical code	n/a	102.1

Listing & Labeling

☐ Appliances require L&L (UMC: L&L or approval by AHJ)	1302.1	301.2
☐ Install per conditions of listing & AMI	1307.1 & 1401.1	303.1
☐ Attach installation & operating instructions to appliance	1307.1	303.1
☐ Fuel-burning factory-applied nameplates must include:	1303.1	307.1–3
• MFR name, model, serial number & listing mark of testing agency		
• Electrical rating (where applicable) in volts, amps & phase		
• Type of fuel and hourly Btu rating		
• Required clearances from combustibles		

Existing Installations

☐ Continued use of existing installations OK if safe & compliant w/ code in effect at time of construction	1202.2	102.2
☐ Owner responsible to maintain existing installations in safe condition & compliant w/ code in effect at time of construction	1202.2	102.3
☐ BO may order reinspections to determine compliance	1202.3	102.3

(IRC Appendix AD contains recommended procedures for safety inspection of existing appliances. The UMC covers procedures for placing gas-fired equipment into operation, and UMC Appendix E addresses sustainable practices.)

MATERIALS & TESTS

Materials	**21 IRC**	**21 UPC**
☐ Plumbing materials must be listed (3rd party certified)	2609.4	301.2
☐ Install components in accordance w/ listing and AMI	2609.2	309.4
☐ Each pipe, fitting, trap, fixture, material & device requires MFR identification & any markings required by applicable standards EXC	2609.1	301.2.1
• Field-cut nipples	2609.1	301.2.1X
• Identification may be on packaging for small fittings	2609.1X	n/a
☐ Locate so as to prevent contamination in flood zones	2601.3	301.4

Water Supply Testing

☐ Test all piping before cover, concealment, or use	2503.2	105.2.1.1&2
☐ Water pipe test under working pressure 15 minutes EXC	2503.7	609.4
• 50 psi air for other than plastic pipe	2503.7	609.4
• Air for PEX where specifically AMI & approved	2503.7X	609.4X
☐ Water for testing must be from potable water source	2503.7	609.4
☐ RP devices tested at installation & annually	2503.8.2	603.4.2

DWV System Testing

☐ Rough-in piping to be tested by one of the following:		
• Water test: 10-ft. head for 15 minutes OR	2503.5.1	712.2
• Air test: 5 psi (10-in. mercury column) for 15 minutes (IRC: No pressurized air testing of plastic piping systems)	2503.5.1	712.3
• Vacuum test: Hold negative 5 psi for 15 minutes	2503.5.1[1]	n/a
☐ Building sewer: Insert test plug at point of connection w/ public sewer; 10-ft. head for 15 minutes (UPC: or air test)	2503.4	723.1
☐ UPC: No air testing of plastic sewer pipe	n/a	723.1
☐ Finished plumbing: Fill each drain, inspect traps	2503.5.2(1)	712.1

PIPE PROTECTION

Freezing Protection	**21 IRC**	**21 UPC**
☐ Pipes subject to freezing require protection (insulation or heat), e.g., outside buildings, attics, crawlspaces, or exterior walls	2603.5	312.6
☐ Applies to water, soil & waste pipes	2603.5 & 3001.2	312.6

Framing & Pipe Protection

☐ Walls, ceilings, etc., to be left in safe structural condition	2603.1	312.11
☐ Drilling & notching per **F1** & **T3,4**	2603.2	312.11
☐ Steel-plate protection for other than CI or Zi steel in notches or holes < 1¼ in. to face of framing (UPC: ≤ 1 in.) **F2**	2603.2.1	312.9
☐ Extend plates min. 2 in. above & below sole & top plates	2603.2.1	n/a
☐ Protection min. 1½ in. beyond outside diameter of pipe **F2**	n/a	312.9
☐ Protective plates min. 16 gauge (UPC: 18 gauge)	2603.2.1	312.9

1. Vacuum test is a newly recognized method in this edition.

TABLE 3 — NOTCHING & BORING JOISTS ◆ 502.8.1

Nominal[A] Dimension Joist or Girder	Max. Diameter Bored Hole	Max. Notch Length	Max. Notch Depth Outer ⅓	Max. Depth End Notch
6	1½ in.	1 13/16 in.	⅞ in.	1⅜ in.
8	2⅜ in.	2⅜ in.	1 3/16 in.	1 13/16 in.
10	3 1/16 in.	3 1/16 in.	1½ in.	2 5/16 in.
12	3¾ in.	3¾ in.	1⅞ in.	2 13/16 in.

A. Table numbers based on actual sizes of nominal dimension lumber: typically 5½, 7¼, 9¼ & 11¼.

FIG. 1 Notching & Boring Joists, Girders & Rafters

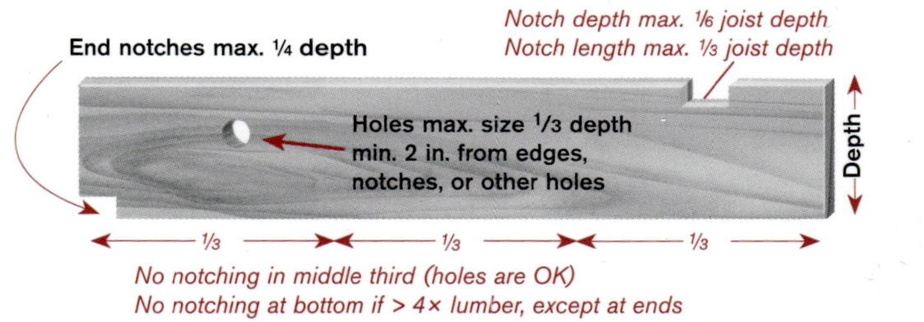

TABLE 4 — MAX. NOTCHES & BORED HOLES IN STUDS ◆ 602.6

Wall Type	Exterior or Bearing Wall		Nonbearing Walls		
	2×4[A]	2×6[A]	2×3[A]	2×4[A]	2×6[A]
Notches	⅞ in.	1⅜ in.	1 in.	1⅜ in.	2 3/16 in.
Holes[B]	1⅜ in.	2 3/16 in.	1½ in.	2⅛ in.	3¼ in.

A. Nominal sizes: table numbers based on actual sizes (½ in. less each dimension).
B. Holes min. ⅝ in. from edge of stud.

FIG. 2

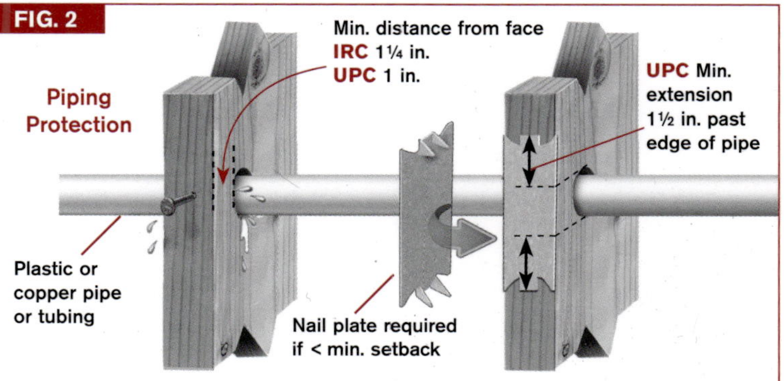

PIPE SUPPORT

Pipe Support	**21 IRC**	**21 UPC**
☐ Hangers must ensure alignment, prevent sagging & allow for expansion (no wires or metal straps contacting plastic pipe) **F3**	2605.1	313.2&4
☐ Isolate dissimilar materials, e.g., Zi hangers & Cu pipe	2605.1	313.2
☐ Max. support intervals for water & DWV piping **T5,6**	2605.1	313.3
☐ Min. hanger rod size ⅜ in.	n/a	313.6
☐ Gas pipe support per **T31**	2424.1	313.7

TABLE 5 IRC WATER PIPE HANGER MAX. SPACING ◆ T2605.1

Pipe Material	Horizontal	Vertical
ABS/PVC DWV	4 ft.	10 ft.[A]
Zi steel	12 ft.	15 ft.
CI	5 ft. (10 ft. OK for 10-ft. lengths of pipe)	15 ft.
Cu tubing	6 ft. if ≤ 1¼ in. diameter, 10 ft. if ≥1½ in.	10 ft.
PB	32 in.	4 ft.
PVC[B]	4 ft.	10 ft.[A]
CPVC	3 ft. for ≤1 in. diameter, 4 ft. if ≥1¼ in.	10 ft.[A]
PEX	32 in. if ≤1in. diameter, 4 ft. if ≥1¼ in.	10 ft.[A]
PEX-AL-PEX	32 in.	4 ft.[A]

A. Provide mid-story guides for pipes ≤ 2 in. diameter.
B. Not allowed for water distribution pipe downstream of water service; OK for condensate drain.

FIG. 3

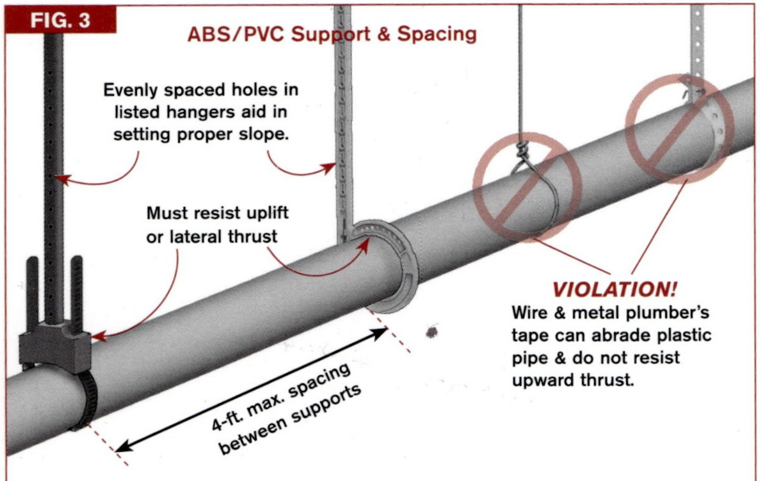

TABLE 6 UPC WATER PIPE HANGER MAX. SPACING ◆ T313.3

Material	Horizontal	Vertical
Schedule 40 PVC & ABS DWV	4 ft.[A]; allow for expansion every 30 ft.	Base & each floor[B]; allow for expansion every 30 ft.
Zi steel	10 ft. if ≤¾ in. diameter, 12 ft. if ≥1 in.	Every other floor, max. 25 ft.
CI lead & oakum	5 ft. (10 ft. if 10-ft. lengths used)[A,C,D]	Base & each floor, max. 15 ft.
CI hubless	Every other joint. If joints > 4 ft. apart, support each joint[A,C,D]	Base & each floor, max. 15 ft.
Cu tubing	6 ft. if ≤1½ in. diameter, 10 ft. if ≥2 in.	Each floor, max. 10 ft.
PVC water distribution[E]	4 ft.[A]; allow for expansion every 30 ft.	Base & each floor[B]; allow for expansion every 30 ft.
CPVC	3 ft. for ≤ 1 in. diameter, 4 ft. if ≥1¼ in.	Base & each floor[B]
PEX	32 in. if ≤1in. diameter, 4 ft. if ≥1¼ in.	Base & each floor[B]
PEX-AL-PEX	98 in.[F]	Base & each floor[B]

A. Support at each horizontal branch connection.
B. Provide mid-story guides.
C. Supports to be within 18 in. of joint. Hangers not to be placed directly on couplings.
D. Brace at max. 40-ft. intervals to prevent horizontal movement.
E. Not allowed for water distribution pipe downstream of water service; OK for condensate drain.
F. MFRs require closer support spacing.

UNDERGROUND INSTALLATION

Pipes in soil must be supported for their entire length. This is especially important for drain and sewer piping to prevent sags that could cause effluent to be trapped and lead to blockage. Smooth, self-compacting backfill such as sand or pea gravel is typically used to achieve this. Pipes must be protected from sharp rocks or other debris in both the bedding and backfill. Piping encased in or passing through concrete requires protection and must not be subjected to structural loads.

Trenching & Backfilling	**21 IRC**	**21 UPC**
☐ Smooth continuous bearing support for pipes in trench	2604.1	313.5
☐ Pipe supported on firm bed, no contact w/sharp objects	2604.1	314.4
☐ No rocks or debris in first 12 in. of backfill over pipe	2604.3	314.4
☐ Backfill in thin layers (max. 6 in.) & tamp in place **F4**	2604.3	314.4
☐ No backhoe or grader until 12 in. of tamped earth **F4**	2604.3	314.4
☐ Parallel trenches not under footings (within 45°) **F5**	2604.4	314.1
☐ Water service min. 12 in. below finished grade **F8**	2603.5	609.1
☐ Plastic underground water service requires #14 blue-insulated tracer wire terminating above ground at each end	n/a	604.10.1
☐ Water pipe min. 6 in. (UPC: 12 in.) below frost line	2603.5	609.1
☐ Building sewer depth per BO & utility (UPC: min. 1 ft.)	2603.5.1	718.3
☐ Plastic drain pipe trench width = pipe O.D. + 16 in. **F4**	n/a	314.4.1

FIG. 4

Plastic Pipes in Trench

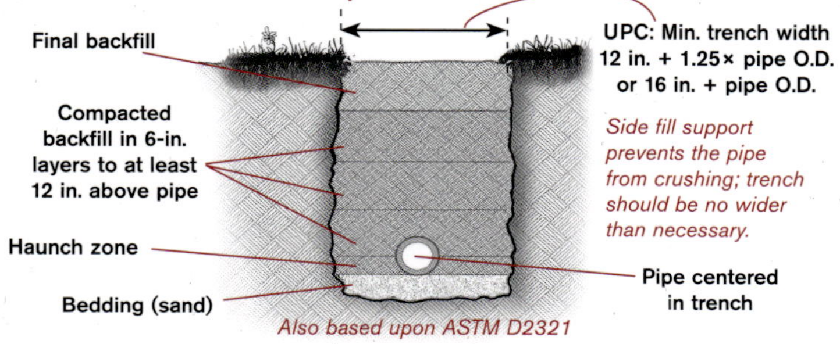

FIG. 5

Pipes Parallel to Footings

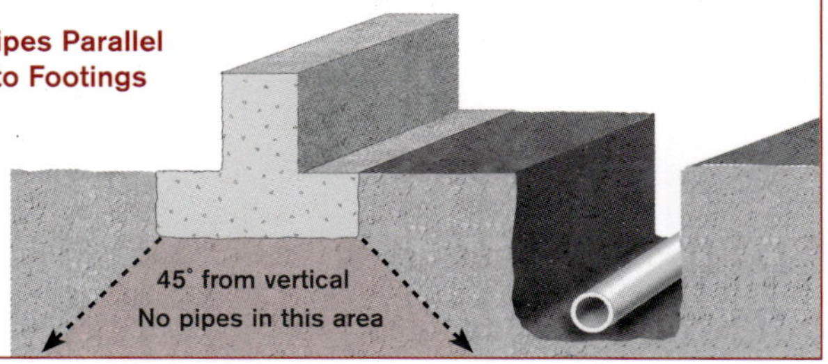

Piping in Concrete or Masonry

	21 IRC	21 UPC
☐ No piping directly embedded in concrete or masonry	n/a	312.2
☐ Protective sheath for piping embedded in corrosive soil	2603.3	n/a
☐ Sheath min. 8-mil plastic & allow for movement	2603.3	n/a
☐ Sleeve required to prevent structural load on pipes through foundation walls or under footings EXC **F6**	2603.4	312.10.1
• Not required for bored or drilled openings	n/a	312.10X
☐ Sleeve min. 2 sizes > pipe through foundation **F6**	2603.4	n/a
☐ Seal annular spaces between pipes & sleeves **F6**	2606.1	312.10.2

FIG. 6

Pipes through Foundation Wall

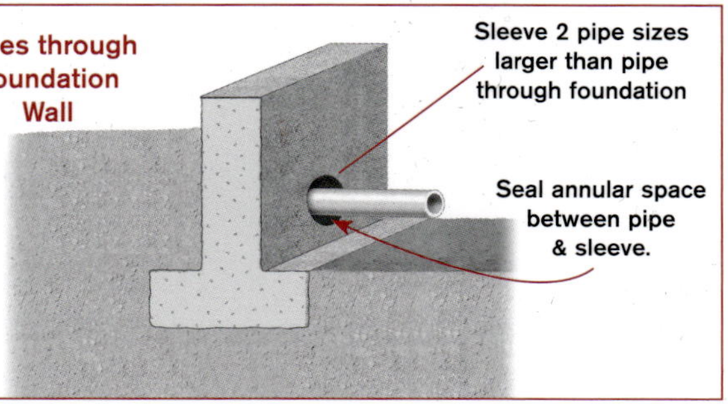

Piping in Common Trench

	21 IRC	21 UPC
☐ Water & sewer OK in same trench if sewer material is a type approved for use inside the building **F7**	2906.4.1	609.2
☐ Water & sewer pipes separated min. 5 ft. if sewer materials not approved for use within the building **F7** EXC	2906.4.1	local
• Water pipe min. 12 in. above perpendicular sewer	2906.4.1	609.2
• Water pipe <12 in. above perpendicular sewer & sleeved min. 5 ft. each direction over sewer pipe crossing	2906.4.1	local
• Water min. 12 in. above parallel sewer pipe **F8**	2906.4.1	609.2
• UPC: On shelf providing 12-in. horizontal separation	n/a	609.2

FIG. 7

Separate & Common Trenches

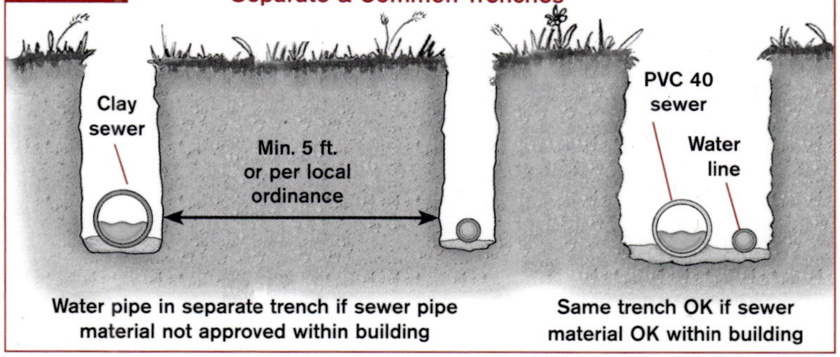

FIG. 8

Water & Sewer Separation

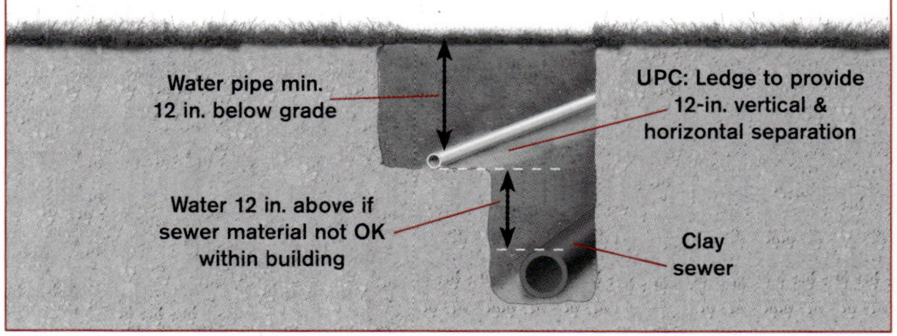

Piping under Concrete Slabs

	21 IRC	21 UPC
☐ Galvanized piping not permitted	3002.1	609.3(1)
☐ Cu pipe to be installed w/o joints where possible	n/a	609.3(2)
☐ Cu pipe joints brazed not soldered	n/a	609.3(2)
☐ All piping tested before casting concrete	109.1.2	105.1

DRAINAGE

Drainage pipe and fittings must have a smooth interior waterway with no ledges or obstructions to flow. Plastics and no-hub cast iron are the most common materials used today. Because of the sound-dampening properties of cast iron, it is sometimes referred to as quiet pipe. Drain piping must have the minimum slopes shown below and should not have excessive slope. If pipes are pitched too steeply, the liquids will outrun the solids and possibly leave them behind, resulting in obstructed pipes.

General	21 IRC	21 UPC
☐ Materials per **T7** & applicable standards	3002.1&2	T701.2
☐ Min. pipe slope ¼ in./ft. EXC	3005.3	708.1
• ⅛ in./ft. OK for 3-in. or larger pipe	3005.3	Ø
• ⅛ in./ft. OK for ≥4-in. pipe where necessary if OK by AHJ	n/a	708.1
☐ No drilled or tapped connections (e.g., saddle fitting)	3003.2	310.2
☐ Joints between different materials require mechanical joints either compression or self-sealing type installed AMI	3003.13	705.1&2
☐ CI couplings require metallic shield & center stop **F10**	3003.4.3	705.2.2
☐ ABS & PVC not directly glued together EXC	3003.2 #5	705.1&5
• Transition of building drain to building sewer OK w/ listed transition solvent cement per ASTM D3138	3003.13.4	705.9.4
☐ No reduction in direction of flow EXC **F9,11**	3002.3.1	315.2
• 4-in. × 3-in. water closet bend OK **F9**	3005.1.6	310.5
☐ ABS & PVC prohibited in residential >2 stories	n/a	California

FIG. 9 Closet Bend Reductions

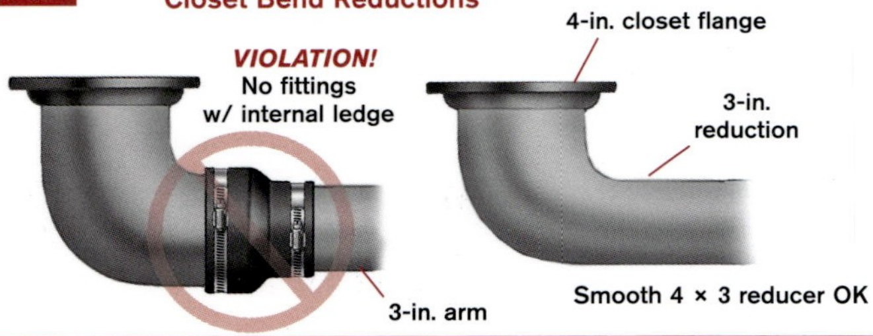

FIG. 10 No-Hub Cast Iron

No-hub cast-iron pipe must be butted tightly to center stop of hub to ensure a smooth interior waterway.

TABLE 7 DWV MATERIALS ◆ IRC T3002.1 & 2 & UPC T701.2

Material	IRC			UPC		
	Above Ground	Below Ground	Building Sewer	Above Ground	Below Ground	Building Sewer
ABS schedule 40	✔	✔	✔	✔	✔	✔
ABS DR 22 & DR 24	✔	✔	✔	Ø	Ø	Ø
ABS/PVC SDR 35[A] (UPC: PVC Sewer & Drain)	Ø	Ø	✔	Ø	Ø	✔
Cast iron	✔	✔	✔	✔	✔	✔
Cu tubing DWV	✔	✔	Ø	✔	✔	✔
Cu tubing K or L	✔	✔	✔	✔[B]	✔[B]	✔[B]
Cu tubing M	✔	✔	Ø	✔[B]	✔[B]	✔[B]
Galvanized steel[C]	✔	Ø	Ø	✔	Ø	Ø
PVC schedule 40	✔	✔	✔	✔	✔	✔
PVC DR 22 & DR 24	✔	✔	✔	Ø	Ø	Ø
PVC 3.25 in. O.D.	✔	✔	✔	Ø	Ø	Ø
Vitrified clay	Ø	Ø	✔	Ø	Ø	✔

A. The IRC also accepts ABS SDR 42 & PVC SDR 26 & 41 for building sewers.
B. UPC 701.2(4) allows Cu alloy tubing thicker than type DWV, even though not stated in T701.2.
C. Min. 6 in. above soil (UPC).

FIG. 11 Durham Systems

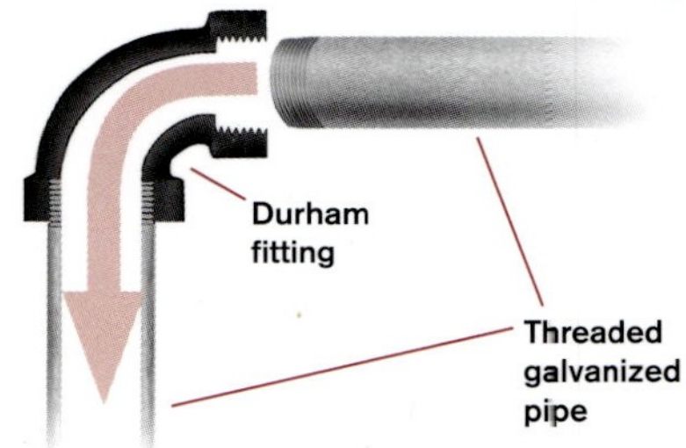

Galvanized pipe is generally intended only for supply piping. If used in drainage with supply-type fittings, the ends of the pipes create a ledge that can accumulate obstructions. Ductile iron Durham fittings have an expanded hub, and the galvanized pipes fully seat into it so the interior waterway is smooth. Galvanized pipes must be kept at least 6 in. clear of soils.

Plastic pipes for drain, waste, and vent (DWV) provide a smooth impervious interior waterway and resistance to corrosion. Durham and no-hub cast iron are more common in commercial construction. In commercial and multifamily installations, pipe must meet standards for flame spread, smoke development, and fire-stopping. Local jurisdictions often restrict the use of plastic piping to residential buildings of a certain type or height. Check with your local building department to learn which rules apply in your area.

*Drain, waste, and vent (DWV) pipe sizes are determined by the number of drainage fixture units (DFUs) that each pipe carries. In the IRC, kitchen, bath, and laundry groups can be sized using **T9**, which allows a smaller number of DFUs than would be calculated if each individual fixture drain were assigned values from **T8**. UPC Appendix C (Alternate Plumbing Systems) provides a similar system. This appendix is not a regular part of the code and its use requires approval by the AHJ. It is most commonly used in multifamily dwellings.*

Drain Pipe Size	**21 IRC**	**21 UPC**
☐ Size traps & arms per DFU loads **T8** ______	T3004.1 & T3201.7	T702.1
☐ Kitchen, bath & laundry group DFUs can be per **T9** ____	T3004.1	Appendix C
☐ Stacks & horizontal branches per DFU loads **T10** ____	3005.4.1	703.1&2
☐ Building drain & horizontal branches directly connected to building drain allowed to be sized per **T11** ________	3005.4.2	n/a
☐ Max. 5 WCs on 3-in. horizontal or vertical drain **T10** _______	n/a	T703.2[2]
☐ Horizontal branches not to connect within 10 pipe diameters downstream of stack base or horizontal offset _________	3005.5	n/a

TABLE 8 — DFUs & TRAP SIZE ◆ IRC T3004.1 & T3201.7 UPC T702.1

Fixture	IRC		UPC	
	DFUs	Trap & trap arm size (in.)	DFUs	Trap & trap arm size (in.)
Bar sink	1	1¼	1	1½
Bathtub (w/ or w/o shower)	2	1½	2	1½
Bidet	1	1¼	1	1¼
Bidet (1½-in. outlet)	1	1½	2	1½
Clothes washer (CW standpipe)	2	2	3	2
Dishwasher (independent drain)	2	1½	2	1½[A]
Floor drain	0	2	0	2
Drinking fountain			0.5	1¼
Kitchen sink (KS)	2[B]	1½	2	1½[A]
Laundry tray (LT)	2	1½	2	1½
Lavatory	1	1¼	1	1¼
Lavatories in sets[C]	2	1½	2	1½
Shower stall	2	1½[D]	2	2[E]
Water closet (toilet) ≤ 1.6 GPF	3	n/a	3	3
Water closet (toilet) > 1.6 GPF	4	n/a	4	3

A. UPC: Min. 2-in. drain.
B. W/ or w/o DW or food-waste grinder.
C. Example would be 2 lavs on one trap.
D. Up to 5.7 gpm; larger traps required for greater flow rates.
E. For a bathtub to shower retrofit, 1½-in. trap & arm permitted for showers ≤ 36 × 60 in.[3]

2. Previous UPC edition limited 3-in. pipe to 4 WCs on vertical drain & 3 on horizontal drain.
3. New in this code edition—allows for simpler process in retrofit of tubs to shower and contradicts the requirement that the shower waste outlet be a min. 2-in. diameter.

TABLE 9 — KITCHEN, LAUNDRY & BATH GROUP DFUs IRC T3004.1 & UPC APPENDIX C T303.3

Group	Description	IRC	UPC[A]
Kitchen	Separate DW & sink	2	n/a
Laundry	CW standpipe + laundry tray	3	n/a
Half bath	1.6 GPF WC + lav	4[B]	3
Full bath	≤1.6 GPF WC, lav, tub w/ or w/o shower	5	5
Full bath	>1.6 GPF WC, lav, tub w/ or w/o shower	6	5½[C]
Multiple bath groups	1 full bath + 1 half bath	7[D]	6[D]

A. Requires adoption of Appendix C or permission of AHJ.
B. Add 1 DFU if WC is >1.6 gal. per flush.
C. Number entered for UPC is for 1.6 GPF pressure-tank WCs.
D. For each additional bath beyond 1½ baths, add 1 DFU per half bath, 2 DFUs per full bath.

TABLE 10 — BRANCH DRAINS & STACKS MAX. DFUS IRC T3005.4.1 & UPC T703.2

Pipe Size		1¼ in.	1½ in.	2 in.	2½ in.	3 in.	4 in.
IRC DFUs	Vertical	1	4	10	20	48	240
	Horizontal	1[A]	3	6	12	20	160
UPC DFUs	Vertical	1	2	16	–	48[B]	256
	Horizontal	1	1	8	–	35[B]	216[C]

A. Limited to a single fixture drain.
B. Based on ¼-in./ft. slope. If ⅛-in./ft. slope, multiply allowable DFUs by a factor of 0.8.

In the IRC, building drain branches are the horizontal pipes that connect directly to the building drain. They can carry drainage from multiple branch drains that are less likely to be all used at the same time. Therefore, the IRC allows the building drain branches and the building sewer a larger number of DFUs than for the upstream branch drains of the same pipe size. The UPC does not have such a rule. Pipe size reductions would require an engineered design approved by the AHJ.

TABLE 11 — IRC MAX. DFUs ON BUILDING DRAIN, BUILDING DRAIN BRANCHES & BUILDING SEWER ◆ IRC T3005.4.2

Pipe Size (in.)	Slope (in. per ft.)		
	⅛	¼	½
1½[A,B]	n/a	Note A	Note A
2[B]	n/a	21	27
2½[B]	n/a	24	31
3	36	42	50
4	180	216	250

A. 1½-in. horizontal branches to building drains limited to 1 pumped fixture (including food-waste grinder) or 2 nonpumped fixtures.
B. Drains <3 in. may not receive discharge from water closets.

Fittings & Changes of Direction

	21 IRC	21 UPC
☐ Direction changes require appropriate fittings **F12–14,T12**	3005.1	706.1
☐ Sanitary tee horizontal to vertical only, not on "back" **F14**	3005.1	706.2
☐ Heel-inlet quarter bends not to serve WC **F13**	3005.1.2	n/a
☐ Heel-inlets OK as dry vents when in vertical position **F13**	3005.1.3	706.2

TABLE 12 APPLICATION OF FITTINGS ◆ IRC T3005.1 & UPC 706

Fitting	Horizontal to Vertical	Vertical to Horizontal	Horizontal to Horizontal
1/16 bend (22.5°)	✔	✔	✔
1/8 bend (45°)	✔	✔	✔
1/6 bend (60°)	✔	IRC ✔ UPC[A]	IRC ✔ • UPC Ø
1/4 bend	✔	IRC[B]	IRC[B] • UPC Ø
Short sweep (hubless CI)	✔	✔[B,C]	IRC[B] • UPC ✔
Long sweep	✔	✔	✔
Sanitary tee[D]	✔	Ø	Ø
Wye	✔	✔	✔
Combo wye & 1/8 bend	✔	✔	✔

A. Branches or offsets permitted if installed in a true vertical position.
B. Allowed for IRC fixture drain ≤2 in. diameter. **F13**
C. 3 in. & larger diameter.
D. See **F12** & comments.

Flow in vertical piping is faster than flow in horizontal piping. The type of fitting for each change of direction relies on that principle. The transition from vertical to horizontal is from fast moving to slow moving, and therefore fittings with greater sweep are required. Horizontal-to-vertical transitions are from slow moving to fast moving and do not require as gradual a bend, so sanitary tees are allowed.

FIG. 12

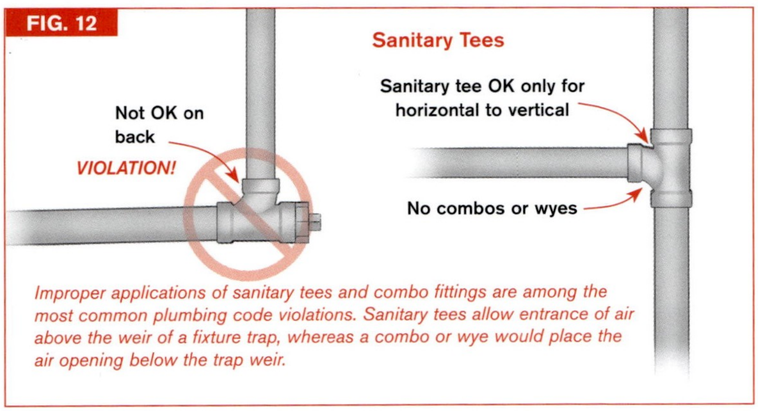

Improper applications of sanitary tees and combo fittings are among the most common plumbing code violations. Sanitary tees allow entrance of air above the weir of a fixture trap, whereas a combo or wye would place the air opening below the trap weir.

FIG. 13

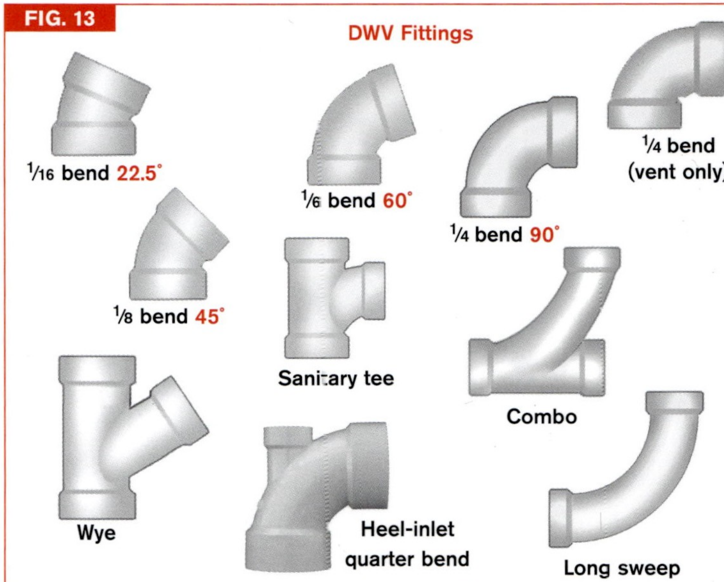

Back-to-Back Horizontal-to-Vertical Fittings

	21 IRC	21 UPC
☐ Permitted if constructed so one inlet does not readily discharge into opposing inlet **F14**	3005.1.1	706.2
☐ Permitted only if opposing connections same size & from similar fixtures or fixture groups	3005.1.1	n/a
☐ Double sanitary tee barrel min. 2 sizes larger than inlets **F14**	n/a	706.2
☐ Double sanitary tee may not receive discharge of back-to-back WCs or pumped fixtures EXC	3005.1.1	n/a
• Horizontal developed length of fixture drains ≥18 in.	3005.1.1X	n/a

FIG. 14

Drains Entering at Same Level

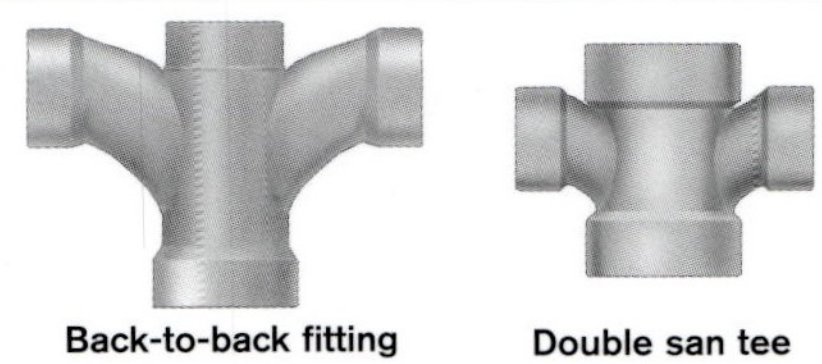

A back-to-back fixture fitting should be used for fixture drains or trap arms entering at the same level. The IRC allows a double sanitary tee to be used for this purpose where the inlets serve similar fixtures and both drains are the same size. The UPC allows double san tees only for drains entering at the same level and into a barrel that is a min. of two sizes larger than the inlets.

Building Drain & Building Sewer

	21 IRC	21 UPC
☐ Building drain = lowest piping carrying all drains in building to the building sewer	202	204.0
☐ IRC: Building sewer begins 30 in. beyond exterior walls	202	n/a
☐ UPC: Building sewer begins min. 2 ft. outside building wall	n/a	204.0
☐ CO required at junction of building drain/sewer or in building drain upstream of junction (IRC: Within 10 ft.) EXC **F15**	3005.2.3	719.1
• If sewer <10 ft. & straight line from building drain	n/a	719.2
☐ Additional building drain COs at max. 100-ft. intervals	3005.2.1	707.4

FIG. 15 2-Way Cleanout

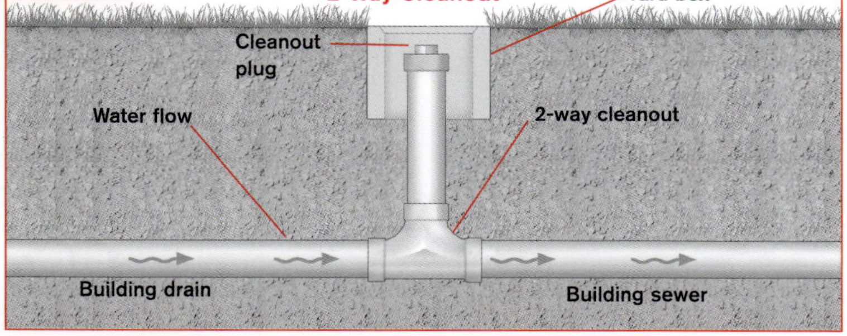

CLEANOUTS

Cleanouts are necessary for clearing drain obstructions and for inspecting the building sewer with a sewer camera. Cleanouts must be accessible, and each code places restrictions and limitations on their location in crawlspaces.

UPC Cleanout Requirements

	21 UPC
☐ COs required to be watertight & gastight	707.3
☐ CO plugs raised square heads or countersunk rectangular slots	707.1
☐ CO plug sizes per **T13**	707.1
☐ Required at upper terminal of all horizontal runs EXC	707.4
• Horizontal runs <5 ft. (unless serving sinks or urinals)	707.4X1
• Horizontal pipes ≤72° from vertical (1/5 bend)	707.4X2
• Pipes (other than building drain) above lowest floor of building	707.4X3
• No upper terminal CO required if 2-way CO at junction of building drain and building sewer **F15**	707.4X4
☐ Required every 100-ft. length or fraction of developed length	707.4
☐ Required for runs w/ aggregate change of direction >135°	707.4
☐ Trap arm bends <90° do not require CO	707.14
☐ Takeoff above flow line unless wye branch or end of line **F16**	707.5
☐ Clearance in front of CO min. 24 in. (18 in. OK if pipe ≤2 in.)	707.9
☐ Underfloor CO must extend above finished floor or outside building if >5 ft. from crawlspace access door or if <18-in. vertical clearance or if passageway to CO <30 in. wide	707.9
☐ Terminate above grade & readily accessible or under approved cover	707.8

IRC Cleanout Requirements

	21 IRC
☐ Required in horizontal drains each change of direction >45° EXC	3005.2.4
• First CO may serve first 40 ft. downstream w/ other changes	3005.2.4
• Not required for horizontal fixture drain w/ nonremovable trap for the section of piping between trap & its vent connection	3005.2.1X
☐ Horizontal drains & building drains max. cleanout interval 100 ft.	3005.2.1
☐ Building sewers <8-in. max. cleanout interval 100 ft.	3005.2.2
☐ CO required at (or within 10 ft. developed length upstream of) junction of building drain & building sewer **F15**	3005.2.3
☐ CO plugs require raised or countersunk sq. head or countersunk slot	3005.2.6
☐ CO same size as drain pipes EXC	3005.2.5
• 4-in. CO OK for piping >4 in.	3005.2.5X
• Removable trap OK 1 size smaller than drain (e.g., kitchen)	3005.2.5X1
☐ CO in crawlspace requires travel path min. 24 in. high	3005.2.10
☐ COs below grade must be extended to grade level **F15**	3005.2.10
☐ Install COs to allow cleaning in direction of flow	3005.2.8
☐ Removal of WC or a fixture trap OK as required CO EXC	3005.2.10.1[4]
• CO for building sewer not to depend on removal of WC	3005.2.3
☐ Pipes ≤6 in. diameter require 18-in. clearance opposite CO	3005.2.9
☐ Not OK to obstruct CO w/ permanent finishes, etc.	3005.2.9
☐ CO openings not OK for new fixtures w/o new CO installed **F16**	3005.2.11

FIG. 16 Adding Drain at Existing Cleanout

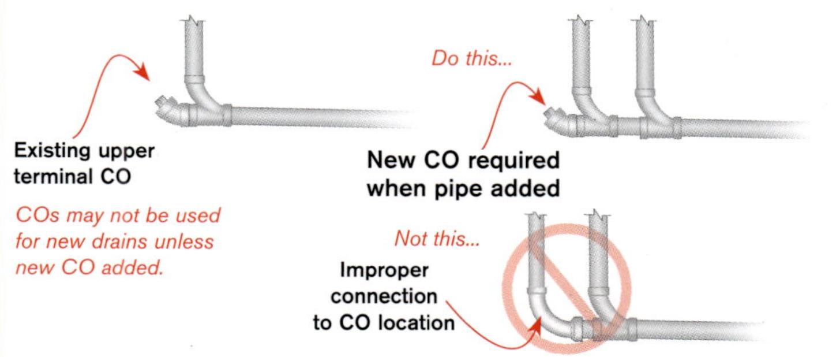

Installing cleanouts above the drain flow line may minimize discharge when removing the cleanout plug.

TABLE 13 — UPC CLEANOUT PLUGS ◆ T707.1

Pipe Size (in.)	Cleanout Size (in.)	Pipe Size (in.)	Cleanout Size (in.)
1½	1½	3	2½
2	1½	4	3½
2½	2½	>4	3½

4. Explicitly allowing a practice that has been common in the field for years.

FIXTURES BELOW SEWER OR MANHOLE

Fixtures above the sewer drain toward it by gravity. Where fixtures or their drains are located below the upstream manhole cover, a sewage backup at the street may result in effluent discharge into the lower fixtures. Backups can be prevented by installing backwater valves on the drain lines that are below the manhole cover. Backwater valves must be accessible for maintenance. Cleanouts that could lead to a backwater valve should be labeled to warn against potential damage to the backwater valve from attempting to run a snake through it.

Fixtures below Upstream Manhole Cover F17	21 IRC	21 UPC
☐ Fixtures below next upstream manhole require backwater valve (measured from FLR in IRC, from floor level in UPC)	3008.1	710.1
☐ Fixtures above elevation of manhole not allowed to discharge through backwater valve EXC	3008.1	710.1
• OK if backwater valve normally open type	3008.2	Ø
• Normally closed type OK in existing buildings	3008.2X	Ø
☐ Backwater valves required to be accessible for service	3008.4	710.6
☐ COs for drains through backwater valve require label	n/a	710.1

FIG. 17

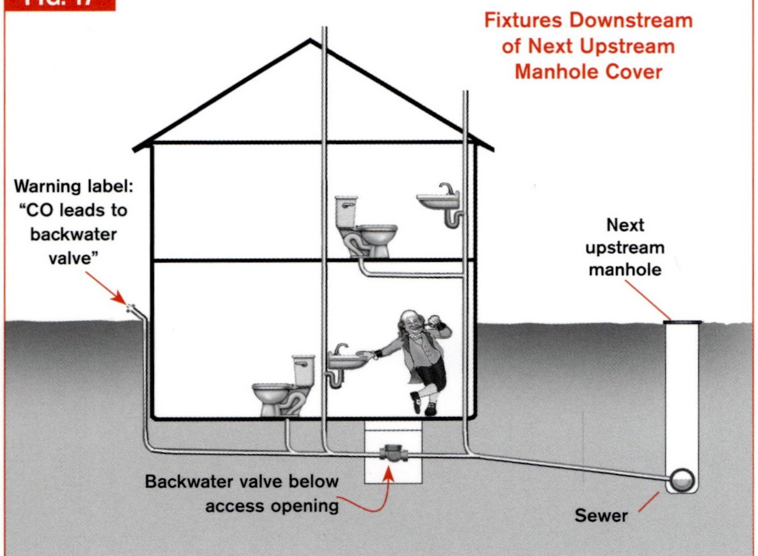

*Fixtures above the sewer but below the next upstream manhole are required to be protected by an accessible backwater valve (a type of check valve) as shown here. The valve protects the building from sewer contamination in the event of a street main backup. The **UPC** also requires a backwater valve for fixtures below the manhole cover of a private sewer system.*

FIG. 18

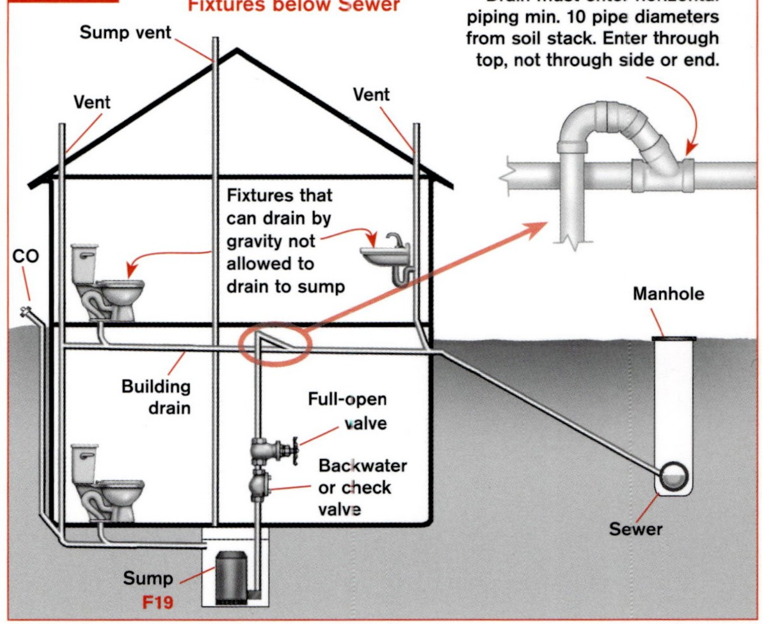

Fixtures below Sewer	21 IRC	21 UPC
☐ Fixtures must drain to sewer by gravity where possible	3007.1	709.1
☐ Sump discharge must be lifted above gravity drain F18	3007.1	710.4
☐ Sump requires gastight removable cover F19	3007.3.2	710.10
☐ Sump pit vented as required for gravity systems	3007.3.2	710.10
☐ Sump vent min. 1¼ in. (UPC: 1½ in.)	T3113.4.1	710.10
☐ Sump effluent level adjustment min. 2 in. below inlet	3007.3.4	710.9
☐ Connect to wye in top of horizontal gravity drain F18	3007.3.5	710.4
☐ Discharge pipe materials must be pressure rated	3007.3.3.1	710.4
☐ Discharge pipe materials Cu, CPVC, CI, PE, or PVC	3007.3.3.2	n/a
☐ Accessible backwater or swing-check valve required on ejector discharge pipe F18,19	3007.2	710.4
☐ Full-open valve at check valve discharge side F18,19	3007.2	710.4
☐ Valve bodies corrosion-resistant metal (no plastic valves)	n/a	710.4
☐ Min. 2-in. discharge piping T14 EXC	3007.6	710.3 #2
• Grinder pumps min. 1¼-in. discharge	3007.6X1	710.12.1
• Macerating toilets min. ¾-in. discharge	3007.6X2	710.13.2
☐ Gravity drains receiving discharge from ejector sized at 1.5 DFU for each gpm of pump (UPC: 2 DFU)	T3004.1	710.5

Sump Pit & Pump

	21 IRC	21 UPC
☐ Sewer ejector pumps must operate automatically	3007.4	710.9
☐ Sump min. 18 in. diameter 24 in. deep **F19**	3007.3.2	n/a
☐ Sump concrete, metal, or other approved materials (IRC: Also tile or plastic; UPC: Metal requires corrosion protection)	3007.3.2	710.8
☐ Min. pump capacity 21 gpm (UPC: 20 gpm)	3007.6	710.3 #1

FIG. 19

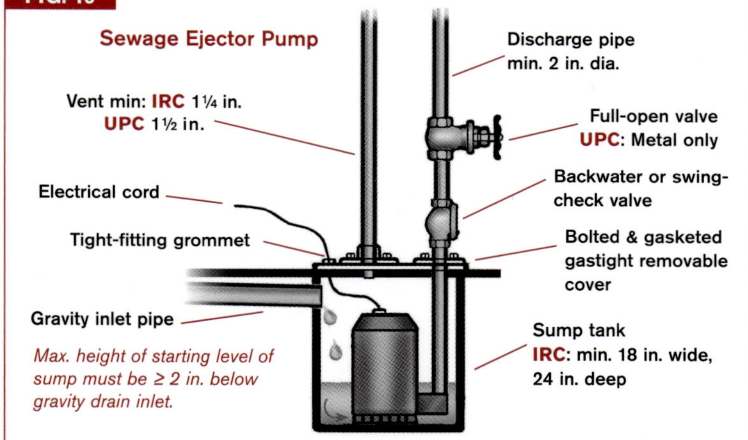

TABLE 14 IRC MINIMUM EJECTOR PUMP CAPACITY ◆ T3007.6

Discharge Pipe Diameter	Pump Capacity
2 in.	21 gpm
2½ in.	30 gpm
3 in.	46 gpm

FIG. 20

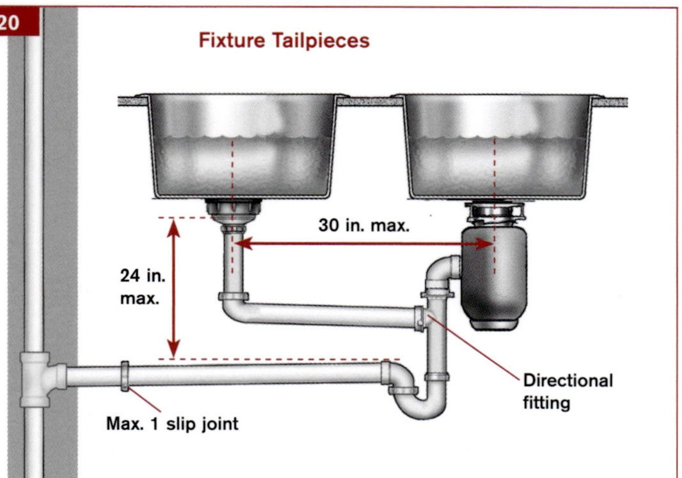

TRAPS, FIXTURE TAILPIECES & TRAP ARMS

Traps provide an air barrier between the contaminated atmosphere of the sewer and the indoor air we breathe. Without a proper trap seal, sewer gases, airborne bacteria, vermin, and other contaminants can enter the living area. If the seal is too shallow, the seal could be lost due to evaporation. If too deep, drainage could be blocked with sludge. Trap arms (fixture drains) must be vented, otherwise the negative pressure created by water moving down the pipe will cause water to be sucked out of the trap and the seal to be lost. Maintaining a proper trap seal is the underlying principle behind the code rules for drainage, traps, and venting.

Traps & Fixture Tailpieces

	21 IRC	21 UPC
☐ Each fixture requires separate trap EXC	3201.6	1001.2
• Fixtures w/ integral traps (toilets)	3201.6X1	1001.2
• 2 or 3 like fixtures (sinks, laundry tubs, or lavs) in same room allowed on single trap at center fixture if fixture outlets ≤30 in. apart **F20**	3201.6X2	1001.2
• Laundry tray (sink) may drain to CW standpipe **F75**	3201.6X3	n/a
☐ Trap seal min. 2 in., max. 4 in. **F21**	3201.2	1005.1
☐ Set traps level & protect from freezing	3201.3	1005.1
☐ Vent opening must be at or above trap weir **F24**	3201.3	1002.4
☐ No "S" traps, bell traps, drum traps, crown-vented traps, or traps w/ moving parts or interior partitions EXC **F23**	3105.3 & 3201.5	1004.1&2
• Lav traps w/ plastic or stainless partitions	3201.5(2)	1004.1
☐ Drum & bottle traps allowed for special conditions	104.11	1004.1
☐ Size trap for fixture per **T8**	3201.7	1003.3
☐ Trap size ≥ fixture tailpiece (UPC: max. one size larger)	3201.7	1003.3
☐ One trap per trap arm, no double traps (in series)	3201.6	1004.1
☐ No corrugated or flexible traps	3201.1	1003.1
☐ Fixture tailpiece max. 24-in. vertical distance EXC **F20**	3201.6	1001.2
• CW standpipes 18–42 in. (UPC: 18–30 in.) **F74**	2706.1.2	804.1
☐ UPC: Tailpiece max. 24-in. total developed length except for the common fixtures joined by continuous waste **F20**	n/a	1001.2
☐ Directional fittings required for continuous wastes from disposer (i.e., wyes, combos, or tees w/baffles) **F20**	2707.1	419.2
☐ Building traps prohibited (UPC: unless required by AHJ)	3201.4	1008.1

FIG. 21

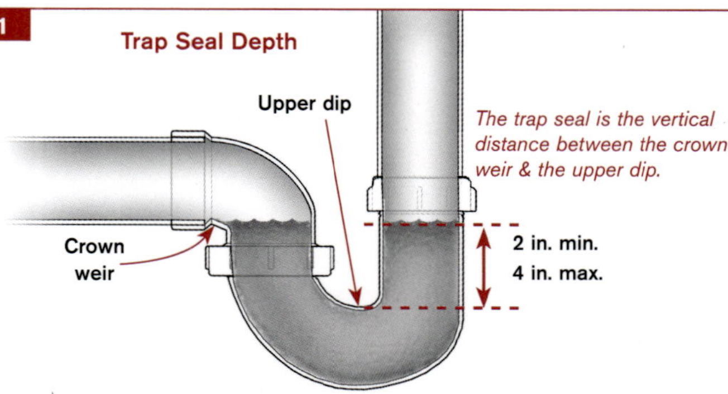

Trap Arms	21 IRC	21 UPC
☐ Trap same size as trap arm	3201.7	1003.3
☐ No crown venting—min. length 2× trap arm diameter **F25**	3105.3	1002.2
☐ Length & slope per table EXC **T15**	3105.1	1002.2
• Trap arm length from WC unlimited (UPC: 6 ft.)	3105.1X	T1002.2
☐ Min. slope ¼ in./ft. (IRC ⅛ in./ft. OK for ≥3-in. pipe)	T3105.1	T1002.2
☐ Total fall of trap arm max. 1 pipe diameter **F24,T15**	3105.2	n/a
☐ Vent connection not below weir of trap (except WCs)	3105.2	1002.4
☐ Only 1 trap permitted on trap arm EXC	n/a	1001.2
• 2 trap arms on same level allowed to join through double-wye fitting to common vent **F29**	3107.2	Ø
☐ Tubing traps require trap adapter **F22**	n/a	1003.2
☐ Max. 1 slip joint allowed on outlet side of trap **F22**	n/a	1003.2
☐ Horizontal direction changes in trap arms per **T12**	n/a	1002.3
☐ CO required if direction change >45° (UPC 90°)	3005.2.4	1002.3
☐ Slip-joint connections required to be accessible **F65**	2704.1	402.10
☐ Access openings min. 12 in. × 12 in. **F65**	2704.1, 3201.1	402.10

FIG. 22

Trap Adapter for Tubing Traps

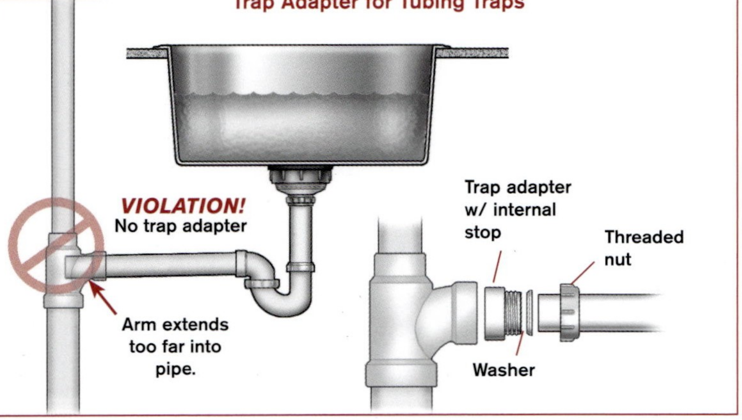

FIG. 23

S Trap

For water to drain in the vertical pipe downstream of the trap, it must have air to prevent a vacuum behind it. If that air comes from the fixture tailpiece, rather than from a vent, the water in the trap also gets siphoned into the drain pipe. That can leave the trap with no water seal to keep out sewer odors & vermin.

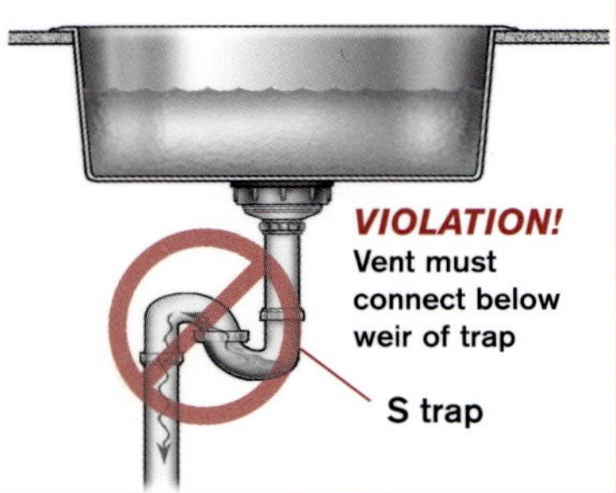

FIG. 24

Trap Arms & Vents

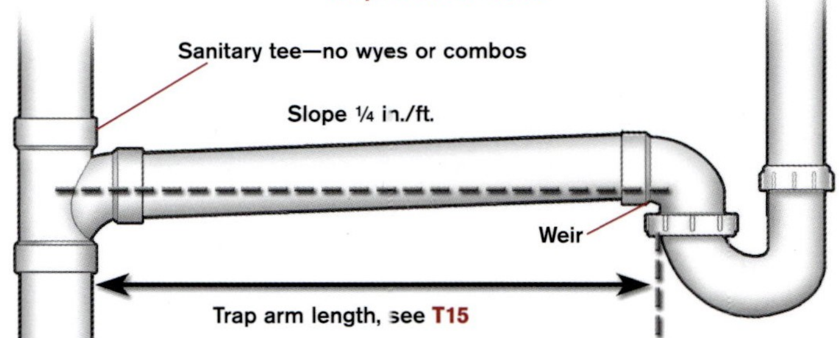

The trap arm must allow air to be admitted above the weir of the trap, indicated by the dotted line in the figure. If the slope or length is excessive, the trap weir would be above the vent opening.

TABLE 15 TRAP ARM DISTANCE TO VENT IRC T3105.1 & UPC T1002.2

Trap Arm Diameter	Min. Arm Length	IRC Max. Arm Length	UPC Max. Arm Length
1¼ in.	2½ in.	5 ft. (60 in.)	2½ ft. (30 in.)
1½ in.	3 in.	6 ft. (72 in.)	3½ ft. (42 in.)
2 in.	4 in.	8 ft. (96 in.)	5 ft. (60 in.)
3 in.[A]	6 in.	12 ft. (144 in.)	6 ft. (72in.)
4 in. or larger[A]	8 in.	16 ft. (192 in.)	10 ft. (120 in.)[B]

A. In the IRC, these arms can have ⅛-in./ft. slope. In the UPC, all arms must slope ¼ in./ft.
B. Max. developed length from a water closet to the vent is 6 ft. in the UPC & unlimited in the IRC.

FIG. 25

Crown Venting

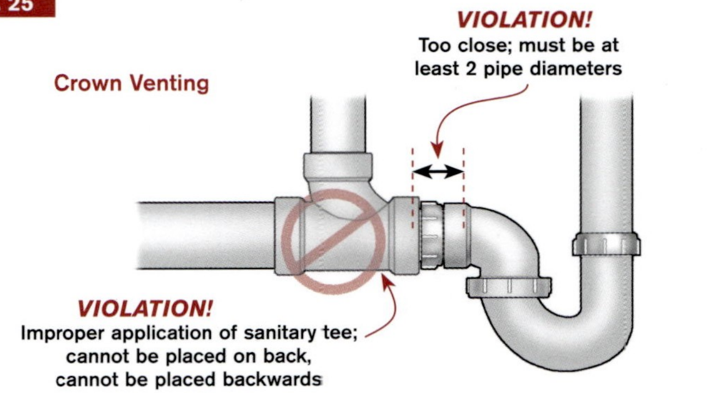

VENTS

Vents prevent atmospheric pressure differences across traps and are essential to maintaining the trap seal. Without vents, the water in the seal could be sucked out, leaving the occupants unprotected from contaminants downstream of the trap. The IRC and UPC have very different approaches to venting.

General

General	21 IRC	21 UPC
☐ All traps require venting	3101.2.1	901.2
☐ Vents must limit trap pressure to max. 1-in. water column	3101.2	901.3
☐ Vent system not to be used for any other purposes	3101.3	local
☐ Vents to horizontal drains must take off above centerline **F28**	3104.3	905.2
☐ Grade & support vents to drain to soil or waste piping	3104.2	905.1
☐ Change direction w/ appropriate fittings **F13,26**	3005.1	903.3
☐ No vent opening below trap weir except toilets	3105.2	905.5
☐ No crown vents: min. 2 pipe diameters from trap **F25**	3105.3	1002.2
☐ Connection to vent stack min. 6 in. above FLR of trapped fixture served by vent	3104.4&5	905.3
☐ Vents not horizontal until min. 6 in. above FLR **F27**	3005.1	905.3
☐ Piping <6 in. above FLR requires drainage-type fittings	3104.5	905.3

Vent Size

Vent Size	21 IRC	21 UPC
☐ Min. size 1¼ in.	3113.1	904.1
☐ Vents min. ½ size of drain served	3113.1	904.1
☐ Size per # of DFUs served & length of vent (UPC: **T16**)	3102.3	904.1
☐ Increase 1 pipe size if developed vent length >40 ft.	3113.1	n/a
☐ Increase 1 pipe size if >⅓ of vent is horizontal	n/a	904.2X
☐ Waste stack vent same size as waste stack	3109.3	n/a
☐ Total area of vents ≥ size of building sewer **T19**	n/a	904.1
☐ Vents for fixtures discharging through pumps, ejectors, or backwater valves do not count towards min. vent requirement	n/a	904.1

TABLE 16 — UPC VENT SIZE AND LENGTH ◆ T703.2

Pipe Size (in.)	1¼	1½	2	3	4
Max. DFUs	1	8[A]	24	84	256
Max. Length (ft.)[B]	45	60	120	212	300

A. Not applicable to water closets, which must have a min. 2-in. vent.
B. Max. horizontal length is ⅓ of total length. If a minimum-sized vent is increased one size, the length limitations of this table do not apply.

Common Vent

Common Vent	21 IRC	21 UPC
☐ Common vent only OK for fixtures on same floor level	3107.1	905.6
☐ Max. 2 traps or trapped fixtures to vertical common vent	3107.1	905.6
☐ Vent connection OK downstream of fixture drains **F29**	3107.2	Ø
☐ Size common vent per DFUs of upper drain **F30,T17**	3107.3	n/a
☐ Upper fixture cannot be WC	3107.3	n/a
☐ UPC: Only allowed to double-inlet fitting **F14**	n/a	905.6

FIG. 26

Branch Vents

Two or more individual vents may combine to form a branch vent, thereby reducing the number of penetrations through the roof. Horizontal vent piping must be graded to drain by gravity to the drain served.

*In the IRC, vents must be at least half the size of the drain they are venting. In the UPC, all vents are sized per **T16** based on their DFUs.*

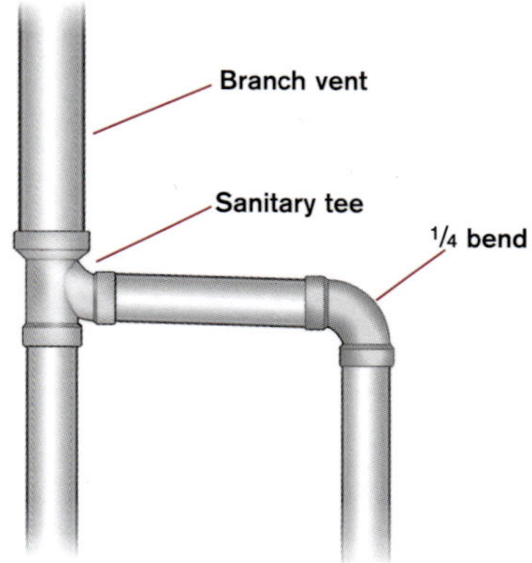

FIG. 27

Vent Offsets

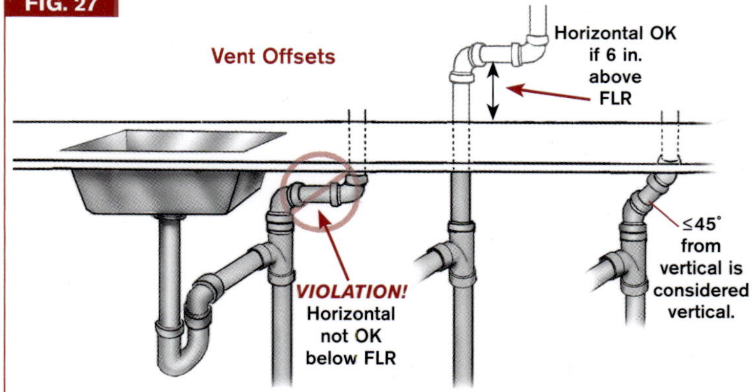

Do not combine vents until 6 in. above highest FLR.

FIG. 28

Vent Takeoff above Centerline

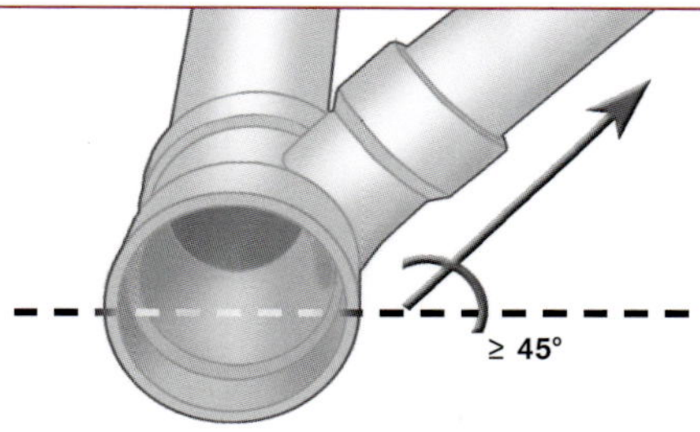

Vent takeoffs must be at least 45° from horizontal. A direct horizontal takeoff could be blocked by effluent in the pipe, which would make the vent ineffective.

FIG. 29

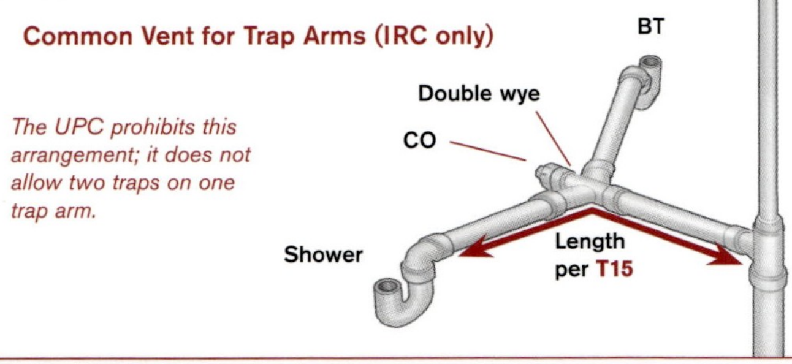

TABLE 17	IRC COMMON VENT SIZES ◆ T3107.3
Pipe Size (in.)	**Max. Discharge from Upper Fixture Drain (Fixture Units)[A]**
1½	1
2	4
2½ to 3	6

A. In the UPC, the vertical wet vent is one size larger than the required size of the drain for the upper fixture drain, as well as a min. of 2 in.

Vertical Wet Venting

	21 IRC	21 UPC
☐ All wet-vented fixtures to be on same story	3108.4	908.1
☐ IRC: Size to # of DFUs discharging to wet vent **T18**	3108.3	n/a
☐ UPC: 1 pipe size larger than waste & min. 2 in. **F31**	n/a	908.1.1
☐ WC fixture drains must connect at same elevation	3108.4	908.1
☐ Each fixture drain connects independently to wet vent	3108.4	n/a
☐ UPC: Limited to trap arms of 1-DFU & 2-DFU fixtures	n/a	908.1
☐ Max. 4 fixtures, max. 6 ft. developed length of wet vent **F31**	n/a	908.1

FIG. 30

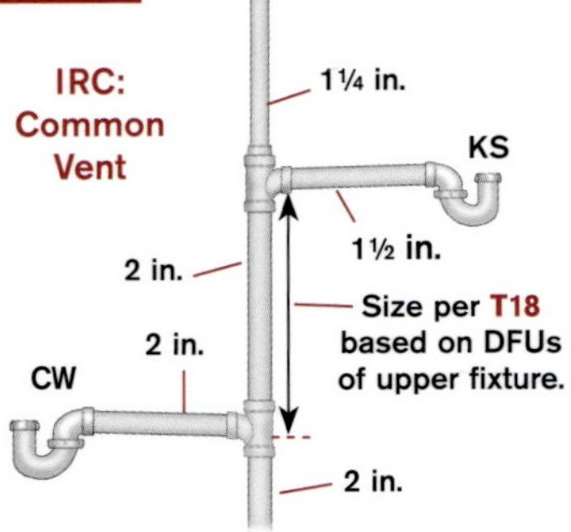

FIG. 31

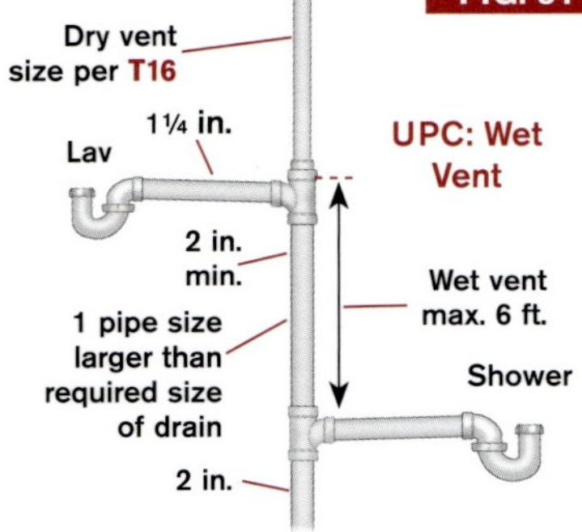

Horizontal Wet Venting

	21 IRC	21 UPC
☐ Horizontal wet venting OK for combinations of fixtures within 2 bath groups (UPC: 1 group) on same floor **F32,33**	3108.1	908.2
☐ Fixtures within the bathroom group(s) only ones allowed on horizontal wet vent (no other upstream fixtures)	3108.1	908.2.5
☐ UPC: WC must be downstream of other drain connections	n/a	908.2.4
☐ Dry vent connection not from floor drain vent	3108.2.1	908.2.1
☐ Dry vent from individual or common vent from any other bath group fixture, if WC drain must connect horizontal	3108.2.1	Ø
☐ Dry vent only individual vent from shower, tub, lav, or bidet	n/a	908.2.1
☐ Max. 1 wet-vented fixture upstream of dry vent connection	3108.2.1	908.2.1
☐ Max. trap arm length from trap weir to wet vent **T15**	3108.5	908.2.3
☐ Size based on DFUs into wet vent **T18**	3108.3	908.2.2

FIG. 32

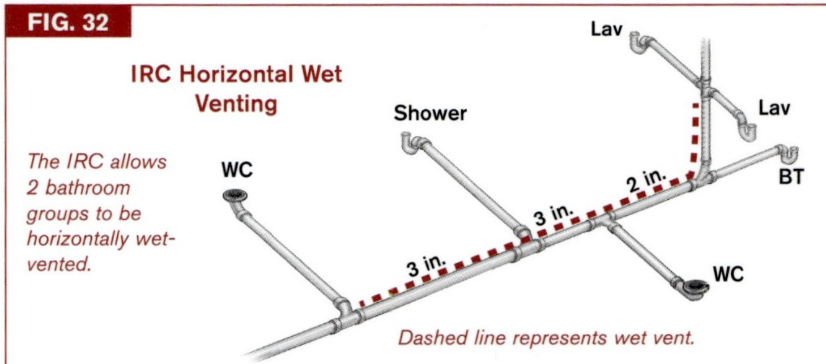

FIG. 33

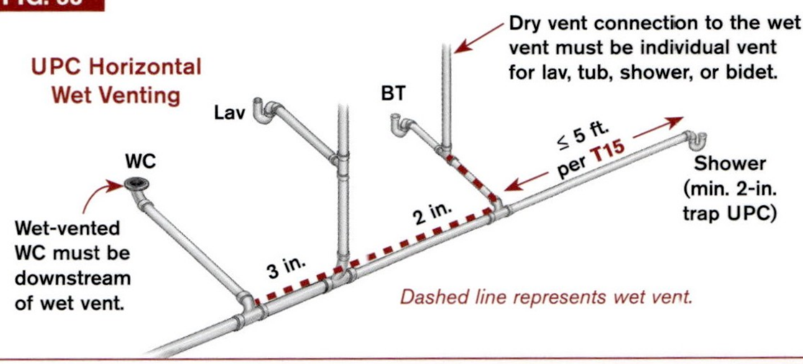

TABLE 18	WET VENT MIN. SIZE ◆ IRC T3108.3 & UPC 908.2.2		
IRC		**UPC**	
# of DFUs	**Min. Pipe Size**	**# of DFUs**	**Min. Pipe Size**
1	1½ in.	1 – 4	2 in.
2 – 4	2 in.	≥ 5	3 in.
5 – 6	2½ in.		
7 – 12	3 in.		
13 – 32	4 in.		

Above sizes are for UPC horizontal wet vents. Vertical wet vents are sized one size larger than the required drain for the fixtures above. The dry vent to wet vent size is per **T16**.

SPECIAL VENTING SYSTEMS

The IRC offers a number of options for island sinks, including air admittance valves, combination waste and vent (CW&V), and the loop vents shown in **F34** and **F35**. The UPC allows only the method in **F34**. In the UPC, a sink in a kitchen peninsula could have a horizontal vent under the countertop and connect it to a vertical vent in the wall if drainage type fittings are used.

Island Sinks	21 IRC	21 UPC
☐ Island venting limited to sinks (disposer OK) & lavs	3112.1	909.1
☐ Island vented w/ drain-type fittings only **F34,35**	3112.3	909.1
☐ Island vent above fixture drain outlet **F35** (UPC: as high as possible **F34**) before returning downward	3112.2	909.1
☐ Lowest part of island vent shall connect full size to a vertical drain or top half of a horizontal drain **F35**	3112.3	n/a
☐ Air admittance valves OK (see AAV section)	3114.3	Ø
☐ COs required in island vents & drains **F34,35**	3112.3	909.1
☐ Connect island vent downstream of fixture drain **F34**	n/a	909.1
☐ Foot vent through wye branch off below-floor vent **F34**	n/a	909.1
☐ CO required in vertical section of foot vent **F34**	n/a	909.1
☐ No upstream fixtures on drain serving island	n/a	909.1

FIG. 34

UPC: Island Sink or Lav

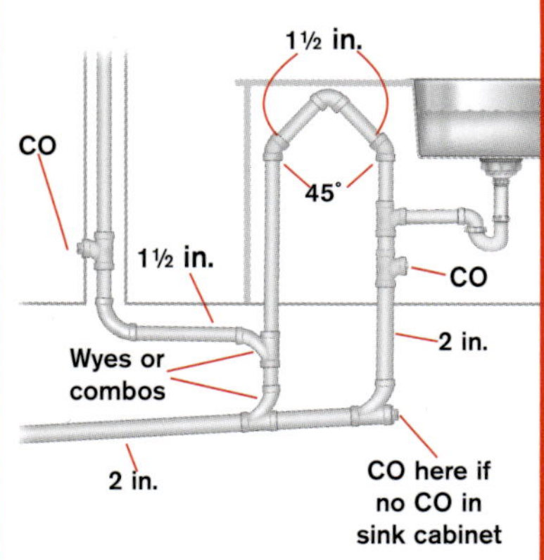

FIG. 35

IRC: Island Sink or Lav

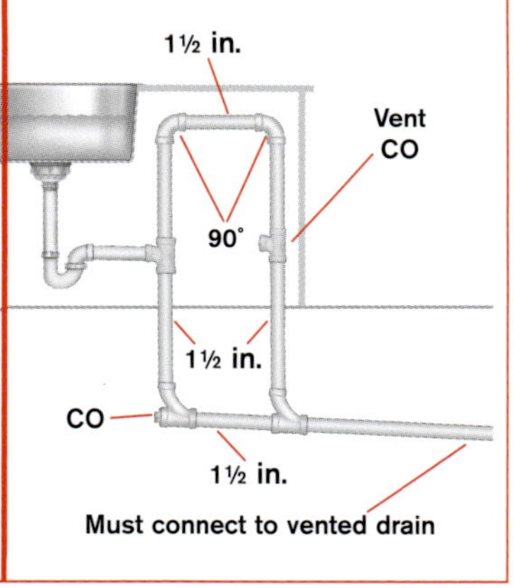

IRC: Combination Waste & Vent (CW&V)	21 IRC
☐ Only allowed for sinks, lavs, floor drains & drinking fountains	3111.1
☐ One vertical pipe (max. 8-ft. length) allowed between fixture drain & horizontal CW&V pipe	3111.2
☐ Max. slope of CW&V piping ½ in./ft., min. slope ¼ in./ft.	3111.2.1
☐ CW&V must connect to dry vent or to vented horizontal drain	3111.2.2
☐ Vent connected to CW&V pipe must rise 6 in. min. above fixture flood level rim before horizontal offsets	3111.2.2

UPC: Combination Waste & Vent (CW&V)	21 UPC
☐ Only allowed where structural conditions prevent conventional system & requires specific advance approval by AHJ	910.1&2
☐ Waste pipe & trap min. 2 pipe sizes larger than otherwise required	910.4
☐ No vertical waste pipe except 1 45° offset allowed in branch lines	910.5

FIG. 36

Air Admittance Valve Operation

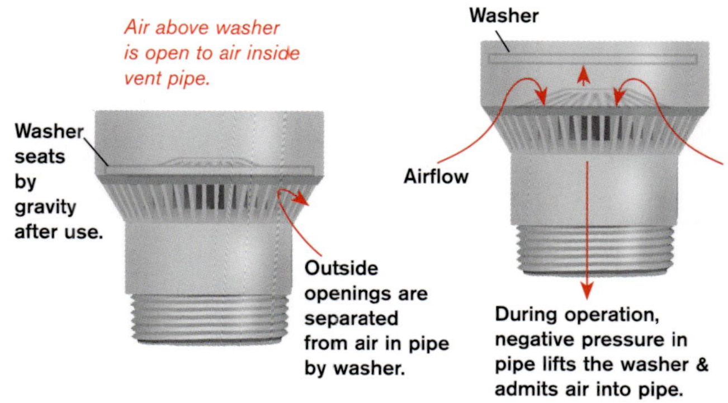

Air admittance valves (AAVs) operate by gravity and have no metal or rubber parts that could corrode or deform. In the IRC they can be used for individual fixtures or for branches. The UPC does not allow AAVs. UPC jurisdictions sometimes accept them on a case-by-case basis under the provisions for Alternate Materials and Methods found in 301.3. If the UPC is the code in your area, be sure to check with your local building department (AHJ) before installing AAVs.

Air Admittance Valves F36	21 IRC
☐ Install after DWV leak test	3114.2
☐ OK at individual, branch, circuit & stack vents	3114.3
☐ Individual & branch type AAV to vent only fixtures on same floor level & that connect to a horizontal branch drain	3114.3
☐ Individual fixture & branch AAV ≥4 in. above branch or fixture drain **F37**	3114.4
☐ Stack-type AAV min. 6 in. above FLR of highest fixture	3114.4
☐ Trap arms same min. & max. length as conventional vent **T15**	3114.4
☐ AAVs terminating in attic min. 6 in. above insulation	3114.4
☐ AAVs must be accessible for service & inspection **F37**	3114.5
☐ Space containing air admittance valve must be ventilated	3114.5
☐ Min. 1 vent to outdoors (UPC: all vents to exterior)	3102.1 & 3114.7
☐ Not OK for sewer ejector pump or tanks	3114.8
☐ Not OK solely to reduce required clearances in **F38**	3114.8

FIG. 37

Air Admittance Valve Locations

An AAV cannot be located inside a stud cavity or other area where not accessible & open to free air.

When AAVs are placed in attics, they must be at least 6 in. above insulation.

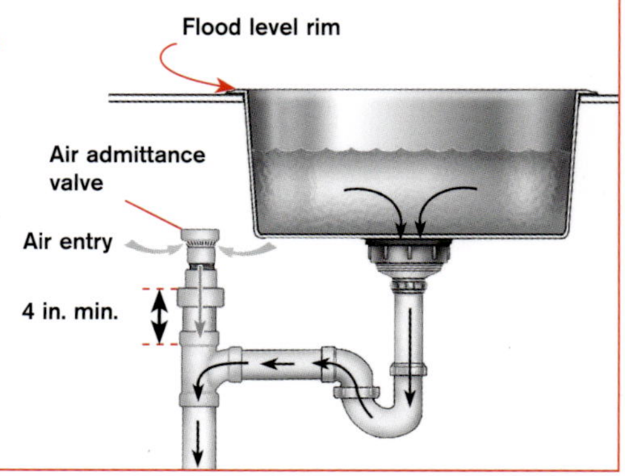

VENT TERMINATION

Vents facilitate drainage flow and protect the seal of traps, thereby preventing sewer odors from entering the living area.

Vent Termination	21 IRC	21 UPC
☐ IRC: sidewall vents allowed; UPC: all vents through roof	3103.1	905.4
☐ IRC: Min. 1 vent must extend outdoors; UPC: all vents	3102.1	906.1
☐ IRC sole vent to exterior not an island fixture vent **F34**	3102.2	n/a
☐ IRC exterior vent min. ½ size of building drain	3102.3 & 3113.1	n/a
☐ Required sum of UPC vents ≥ size of building sewer **T19**	n/a	904.1
☐ Vents through roof min. 6 in. above roof EXC	3103.1.1	906.1&3
• 2 in. OK if vent terminates under solar panel	3103.1.3	Ø
☐ Protect exposed plastic pipe above roof w/ latex paint	MFR	906.1
☐ Min. 7 ft. above roof decks (UPC: or if within 10 ft.)	3103.1.2	906.3
☐ Flagpoling of vents prohibited except where roof is used for parking or assembly purposes	n/a	906.3
☐ Min. 12 in. horizontal from adjacent vertical surfaces	n/a	906.1
☐ Min. 3 ft. above openings within 10 ft. EXC **F38**	3103.5	906.2
• OK if ≥4 ft. below building openings **F38**	3103.5	n/a
☐ Roof termination min. 3-ft. distance from LL other than street	n/a	906.2
☐ IRC sidewall terminations min. 10 ft. from LL, min. 10 ft. from grade within 10 ft., not under vented eaves or soffits	3103.1.4	Ø
☐ Sidewall terminations require bird & rodent protection	3103.1.4	Ø
☐ Vent termination not to be used for any other purpose	3103.4	n/a
☐ Provide flashing for roof penetrations	2607.1 & 3103.3	906.5
☐ If flashing tucked into vent, max. reduction 1 pipe size	2607.1	n/a

Snow or Frost Closure (Design Temperature <0°F)

Item	21 IRC	21 UPC
☐ Min. 3-in.-diameter termination (UPC: min. 2 in.)	3103.2	906.7
☐ Increase vent diameter min. 1 ft. inside thermal envelope	3103.2	906.7
☐ Terminate min. 6 in. above snow line (UPC: 10 in. above roof or per AHJ)	3103.1.1	906.7

FIG. 38

Vent Termination Clearances to Building Openings

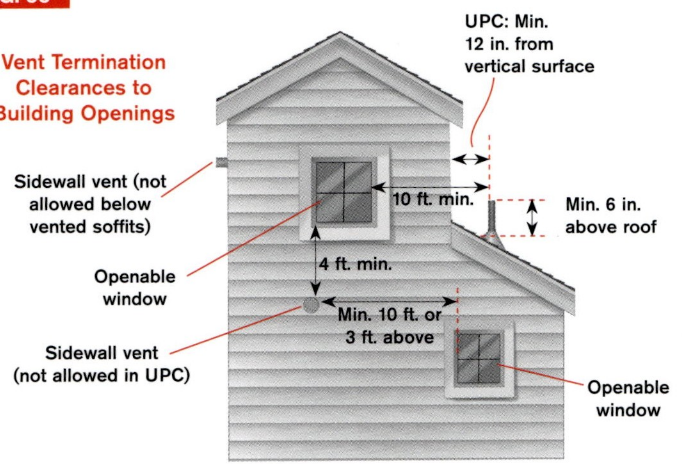

TABLE 19 UPC VENT AREA WORKSHEET ◆ UPC 904.1

Vent Size (in.)	Area (sq. in.)	# Vents	Net Vent Area
1¼	1.23		
1½	1.77		
2	3.14		
3	7.07		
4	12.57		
TOTAL			

Use the fill-in area to the left to determine if the vents through the roof meet the required min. area.

Example:
4-in. bldg. drain = 12.57 sq. in.
One 1¼-in. vent = 1.23 sq. in.
Three 2-in. vents = 9.42 sq. in.
1.23 + 9.42 = 10.65 sq. in.
Thus, more venting through roof is required to meet or exceed the area of the building drain.

*The UPC requirement for the aggregate area of vents being no less than the size of the building drain can be met by having more vents through the roof, or by upsizing where vents are joined as branch vents as in **F26**.*

ON-SITE DISPOSAL SYSTEMS

On-site sewage disposal systems are generally regulated by local environmental health services agencies. Septic systems must be sized and located so as not to impact potable water supply. The areas shown in **T21** are a general planning guide only; local regulations must be followed. Soil percolation testing and the type of construction will determine the actual size. In the IRC, users are directed to use the International Private Sewage Disposal Code (IPSDC). In the UPC, on-site systems are found in Appendix H and are subject to local adoption.

Septic Tanks	21 IPSDC	21 UPC
☐ Locate system on same parcel as building served EXC	101.2	H101.8
• Common systems to comply w/ regs for public systems	101.2	Ø
☐ Tank min. 5 ft. from building	802.8	H101.8
☐ Tank & disposal field min. 5 ft. from lot line	T406.1	T H101.8
☐ Tank min. 25 ft. (50 ft. UPC) from water wells	T802.8	T H101.8
☐ Disposal field min. 50 ft. (UPC: 100 ft.) from water wells	T406.1	T H101.8
☐ Min. tank capacity per **T20**	802.7.1	H201.1
☐ Design to minimize inundation in flood hazard area	303.2	H101.5X

TABLE 20 SEPTIC TANK CAPACITY ◆ IPSDC T802.7.1 UPC H201.1(1)

Bedrooms	Min. Tank Size IPSDC	Min. Tank Size UPC	Max. UPC DFUs
1-2	750	750	15
3	1,000	1,000	20
4	1,200	1,200	25
5	1,425	1,500	33
6	1,650	1,500	33

TABLE 21 1 & 2 FAMILY – MIN. ABSORPTION AREA ◆ IPSCD T603.1

Percolation Class	Percolation Rate[A]	Seepage Trenches[B]	Seepage Beds[B]
1	0 to <10	165	205
2	10 to <30	250	315
3	30 to <45	300	375
4	45 to 60	330	415

A. Minutes required for water to fall 1 in.
B. Sq. ft. per bedroom.

WATER SUPPLY & DISTRIBUTION

Piping must be protected against movement and damage during construction. Plastics typically can tolerate exposure to sunlight for only limited periods, and some types of plastics require flushing procedures to remove solvents prior to use.

Materials	21 IRC	21 UPC
☐ All materials must meet NSF 61 & be approved **T22**	2906.4&5	604.1
☐ Water supply system pipe & fittings max. 8% lead content for non-drinking water applications	2906.2	604.2.1
☐ Drinking water pipe, fittings, joints, valves & fixtures weighted average lead content of max. 0.25%	2906.2.1	604.2
☐ Cu alloy fittings & valves >15% Zi by weight & used w/ plastic piping require resistance to dezincification per NSF 14	2609.3	604.1
☐ Cu pipe markings: K = green, L = blue, M = red	n/a	604.4
☐ Piping or tubing previously used for any purpose other than potable water may not be used for potable water	local	604.4

TABLE 22 WATER PIPE MATERIALS ◆ IRC T2906.4&5 & UPC T604.1

Material	IRC		UPC	
	Service	Distribution	Service	Distribution
ABS	✔	Ø	Ø	Ø
CPVC	✔	✔	✔	✔
CPVC-AL-CPVC	✔	✔	✔	✔
Cu alloy pipe (brass)	✔	✔	✔	✔
Cu tubing	✔	✔	✔	✔
Ductile iron	✔	Ø	✔	✔
Galvanized steel	✔	✔	✔	✔
PE pipe or tubing	✔	Ø	✔	Ø
PE-AL-PE	✔	✔	✔	✔
PE-RT	✔	✔	✔	✔
PEX	✔	✔	✔	✔
PEX-AL-PEX	✔	✔	✔	✔
PEX-AL-HDPE	✔	✔	Ø	Ø
PP	✔	✔	✔	✔
PVC	✔	Ø	✔	Ø

Joints & Connections	21 IRC	21 UPC
☐ Solder joints ASTM B828; solder & flux max. 0.2% lead	2906.15	605.1.4
☐ Flux per ASTM B813 & rated for potable water	2906.15	605.1.4
☐ Joints between dissimilar materials AMI	2906.18	605.16
☐ Cu to Fe use brass (UPC: min. 6 in.) or dielectric fitting	2906.18.1	605.16.1
☐ PVC service to CPVC distribution mechanical fittings or solvent-cement joint per ASTM F493	2906.18.2	605.2.2
☐ CPVC joints AMI, primer must be orange EXC	2906.9.1.2	605.2.2
• One-step cements AMI yellow (IRC: or red)	2906.9.1.2	605.2.2

Joints & Connections (Continued)	21 IRC	21 UPC
☐ Plastic to other piping w/ approved adapter fittings	2906.18.3	605.16.2
☐ Female threaded CPVC fittings w/ male plastic only	local	605.2.3
☐ PVC threaded fittings min. schedule 80	local	605.12.3
☐ Female threaded PVC fittings only OK w/ male plastic	local	605.12.3
☐ Press-connect joints on Cu tubing AMI w/ MFR's tool	2906.19	605.1.3.2
☐ Push-fitting joints OK on CPVC, PEX, PE-RT & Cu tubing that is outside-dimensioned on Cu tubing size, AMI & per ASSE 1061	2906.21	605.9.3
☐ Cu joints in or under concrete slab on grade within building require brazed wrought-copper fittings	local	609.3(2)
☐ Slip joints only on exposed fixture supply	local	605.14
☐ Unions required within 12 in. of WHs, softeners, filters, regulators & similar equipment that may require removal for servicing	local	609.5
☐ Except for necessary valves, intermixing of dissimilar metals only at exposed or accessible locations	local	310.6

Valves		
☐ Accessible main valve required near water entrance	2903.9.1	606.2
☐ Main valve must have bleed orifice or separate drain	2903.9.1	n/a
☐ Additional valve at curb, meter, or lot line per local	2903.9.1	n/a
☐ Main valve must be on discharge side of water meter	local	606.2
☐ Main valve required for each separate building EXC	local	606.2
• SFD & accessory building OK to have only 1 valve	local	606.2
☐ In multiple dwelling units, shutoff valves required each unit	local	606.3
☐ Main & WH valve must be full-open type **F39,56**	2903.9.1&2	606.2
☐ Throttling valves not OK for main & WH **F39**	2903.9.1&2	606.2
☐ Valve required on cold water supply to WHs	2903.9.2	606.2
☐ Required shutoffs must be accessible	2903.9.3	606.6
☐ Hose bibbs subject to freezing require accessible stop & waste valve inside building EXC	2903.10	local
• Frostproof hose bibbs w/ stem into heated space	2903.10X	603.5.7
☐ Valves required each appliance or fixture supply EXC	2903.9.3	606.5
• Tubs & showers (UPC: fixtures w/o slip joints)	2903.9.3	606.5
☐ Valves OK at accessible manifold **F40** if labeled EXC	2903.8.4	606.5
• If manifold in attic, crawlspace, or otherwise not readily accessible, separate shutoff at each individual fixture	n/a	606.5

FIG. 39 Shutoff & Flow Control Valves

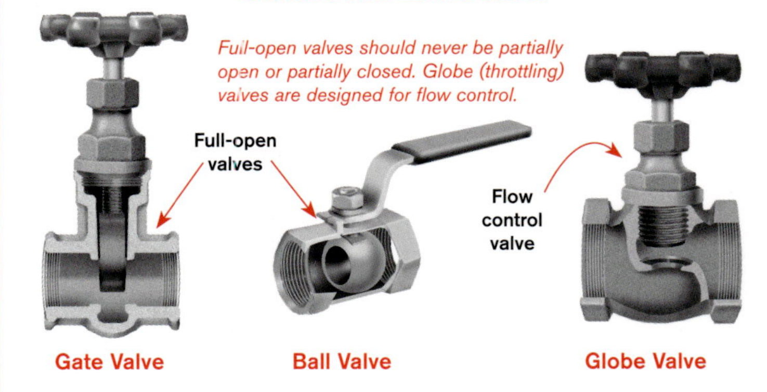

Pipe Installation: General — 21 IRC — 21 UPC

- ☐ PVC not to be exposed to direct sunlight unless max. 24 in. & protected by 0.04-in.-thick tape or other UV protection ____ n/a — 605.12
- ☐ Mechanical ASSE 1010 water hammer arrestors (not air chambers) required near quick-close valves (DW or CW) ________ 2903.5 — 609.11
- ☐ Dead legs must have method of flushing ________ n/a — 309.6[5]
- ☐ Flex connectors accessible (UPC: readily accessible) __ 2906.7 — 604.5
- ☐ Burred ends to be reamed to full bore of pipe or tubing _ 2608.1 — 609.1

PEX Tubing

- ☐ Secure AMI & support per **T5**,**6** ________ 2903.8.3 — 313.3
- ☐ Wrap or sleeve bundles at change of direction ≥45° _ 2903.8.3 — MFR
- ☐ Bend radius AMI **T23** ________ MFR — MFR
- ☐ No PEX within first 18 in. of tank-type WH ________ MFR — 604.13

The Plastic Pipe and Fittings Association and the Plastic Piping Educational Foundation publish helpful design guides and training materials for PEX users. These include the following recommendations:

- Do not store unprotected PEX tubing outdoors and exposed to sunlight.
- Keep tubing & fittings in original packaging until time of installation.
- Minimize sunlight exposure as recommended by MFR.
- Maintain adequate distance between tubing and heat sources, such as light fixtures, flue vents, and heating appliances, so that the pipe surface temperature does not exceed 180°F.
- Supports should allow free tubing movement for thermal expansion.
- Tubing should not be pulled tight; allow slack for expansion/contraction.
- Provide for expansion/contraction using offsets & expansion loops
- Min. bend radius 8× tubing diameter in direction of coil bend. **T23**
- Do not use supports that pinch or cut the tubing.
- Provide support near bends to prevent stress on fittings.
- Do not install within first 18 in. of connections to WH.

TABLE 23 — RECOMMENDED[A] MIN. BEND RADIUS FOR PEX

Tubing Size (in. nominal)	Tubing Outer Diameter (in.)	Bend Radius (in.)
3/8	1/2	4
1/2	5/8	5
3/4	7/8	7
1	1 1/8	9

A: If bending the piping in the opposite direction from which it came off the coil, bending radius should be 3× the value shown in the table.

Pipe & Tubing Size — 21 IRC

- ☐ Min. water service 3/4 in. ________ 2903.7
- ☐ Size of mains, branches & risers determined by demand gpm, available pressure & friction losses from water meter & developed length of pipe including equivalent length of fittings ________ 2903.7
- ☐ Size parallel system per **T25–27** ________ 2903.8.1
- ☐ Size system to provide required flow at peak demand **T24** ________ 2903.1
- ☐ If WH fed from end of cold-water manifold, **F40** manifold one size larger than WH feed ________ 2903.8.2
- ☐ Distribution lines ≤ 60 ft. can be 3/8 in. if meter pressure ≥40 psi ___ 2903.8.2
- ☐ Lines >3/8 in. required AMI for fixtures such as tubs & showers ______ 2903.8.2

5. This new rule prohibits "air chamber" water hammer arresters.

WATER SUPPLY SIZING

Water supply systems can be a "trunk and branch" system or a parallel system. Trunk and branch systems start with a large supply pipe and progressively smaller pipes go toward each branch and fixture outlet. Parallel systems have a manifold **F40** supplying "home runs" to each fixture, or there could be remote manifolds near clusters of fixtures. Each system has its advantages. A trunk and branch system will use less pipe but more fittings. Systems with manifolds generally use PEX tubing. The IRC supply size rules are based on parallel systems, while the UPC methods are based on the trunk and branch system. Each code requires a design that meets minimum required flow rates at the fixture outlets. Both methods start with assigning values of water supply fixture units (WSFUs) to determine the required capacity.

TABLE 24 — MIN. CAPACITIES AT FIXTURE OUTLETS ◆ IRC T2903.1

Fixture Outlet	Flow Rate (gpm)	Flow Pressure (psi)
Bathtub (BT)	4	20
Bidet	2	20
Dishwasher (DW)	2.75	8
Laundry tray (LT)	4	8
Lavatory/Lav (Sink)	0.8	8
Shower (pressure-balancing or thermostatic mixing)[A]	2.5	20
Hose bibb	5	8
Sink	1.75	8
Water closet (WC) (tank type)	3	20
WC (flushometer)	1.6	20
WC (one piece)	6	20

A. Where valve MFR indicates lower flow, use lower value.

TABLE 25 — WSFU FILL-IN TABLE ◆ IRC T2903.6 & UPC T610.3

Fixture	IRC			UPC	#	Extension
	Hot	Cold	Comb.	Comb.[A]		
Bathtub	1.0	1.0	1.4	4.0		
CW	1.0	1.0	1.4	4.0		
DW	1.4	—	1.4	1.5		
Hose bibb	—	2.5	2.5	2.5[B]		
KS	1.0	1.0	1.4	1.5		
Lav	0.5	0.5	0.7	1.0		
LT	1.0	1.0	1.4	1.5		
Shower (per head)	1.0	1.0	1.4	2.0		
Water closet	—	2.2	2.2	2.5		
Laundry group[C]	1.8	1.8	2.5	—		
Kitchen group[D]	1.9	1.0	2.5	—		
Half-bath group[E]	0.5	2.5	2.6	—		
Full-bath group[F]	1.5	2.7	3.6	—		
					Total Demand	

A. UPC fixtures that have both hot & cold connections permitted to use 75% of listed total value of fixture.
B. First hose bibb counts as 2.5 WSFU. Additional hose bibbs add 1.0 WSFU.
C. Laundry group = CW standpipe and LT.
D. Kitchen group = DW and sink w/ or w/o garbage grinder.
E. Half-bath group = water closet and lavatory.
F. Full-bath group = water closet and lavatory and tub (w/ or w/o shower) or shower stall.

*WSFU values for individual fixtures take into account the fact that not all fixtures are used at the same time. The size of system and branch supplies in **T25** has a graduated scale for determining the size of a manifold based on this principle. The next step after determining the total WSFUs is to convert to gpm using **T26**. The manifold can then be sized using **T27**. Refer to the MFR instructions for other considerations of pipe length, pressure, and elevation changes.*

TABLE 26 — CONVERTING WSFUs TO GPM ◆ IRC T2903.6(1)

Load (WSFUs)	Demand (gpm) (flush tanks)[A]	Demand (gpm) (flushometer)[A]	Load (WSFUs)	Demand (gpm) (flush tanks)[A]	Demand (gpm) (flushometer)[A]
1	3.0	–	14	17.0	30.2
2	5.0	–	15	17.5	31.0
3	6.5	–	16	18.0	31.8
4	8.0	–	17	18.4	32.6
5	9.4	15.0	18	18.8	33.4
6	10.7	17.4	19	19.2	34.2
7	11.8	19.8	20	19.6	35.0
8	12.8	22.2	25	21.5	38.0
9	13.7	24.6	30	23.3	42.0
10	14.6	27.0	35	24.9	44.0
11	15.4	27.8	40	26.3	46.0
12	16.0	28.6	45	27.7	48.0
13	16.5	29.4	50	29.1	50.0

A. Choose demand from appropriate column based on whether predominately flush tanks or flushometer.

TABLE 27 — MANIFOLD SIZING ◆ IRC T2903.8.1

Inlet Pipe Size (in.)	Max. GPM Plastic[A]	Max. GPM Metal[B]
3/4	17	11
1	29	20
1 1/4	46	31
1 1/2	66	44

A. Based on velocity limitation of 12 ft./second.
B. Based on velocity limitation of 8 ft./second.

FIG. 40

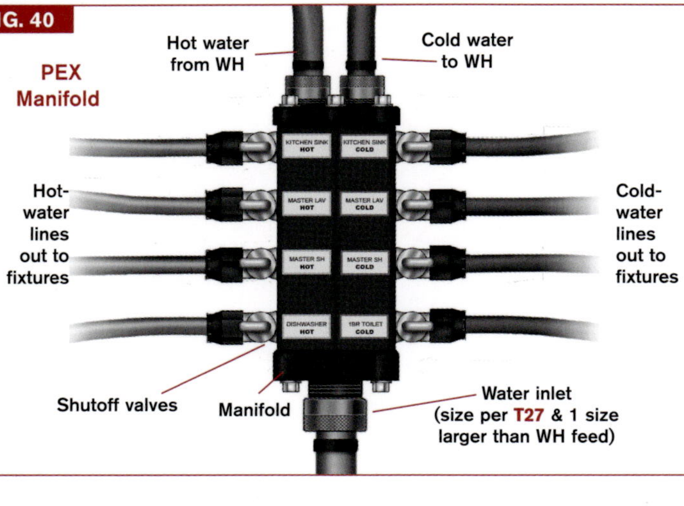

*The procedure for sizing in the UPC is based on the number of fixture units served by each section of piping, including mains and branches. A tree diagram showing the WSFUs on each section of piping can then be filled in with the appropriate sizes from **T29**. A similar method is also included in IRC Appendix AP.*

TABLE 28 — METER & PIPE SIZING WORKSHEET ◆ UPC 610.4

1. Determine fixture unit demand (total from **T25**).	
2. Determine min. daily static pressure at meter or source.	
3. Subtract (add) ½ lb. pressure per ft. of rise (fall).	
4. Deduct pressure losses for filters, regulators, etc.	
5. Find pressure range group in **T29**.	
6. Find column for developed length to most remote fixture.	
7. Find row meeting fixture unit demand (total from **T25**).	
8. Find required meter & pipe size in left column of **T29**.	

NOTE: The same procedure can be used for branches.

TABLE 29 — WATER SUPPLY SIZING ◆ UPC T610.4

Meter (in.)	Supply (in.)	Max. Fixture Units Allowed Per Length of Pipe					
30-45 PSI		40 ft.	60 ft.	80 ft.	100 ft.	150 ft.	200 ft.
3/4	1/2A	6	5	4	3	2	1
3/4	3/4	16	16	14	12	9	6
3/4	1	29	25	23	21	17	15
1	1	36	31	27	25	20	17
3/4	1 1/4	36	33	31	28	24	23
1	1 1/4	54	47	42	38	32	28
1 1/2	1 1/4	78	68	57	48	38	32
1	1 1/2	85	84	79	65	56	48
1 1/2	1 1/2	150	124	105	91	70	57
2	1 1/2	151	129	129	110	80	64
46-60 PSI		40 ft.	60 ft.	80 ft.	100 ft.	150 ft.	200 ft.
3/4	1/2A	7	7	6	5	4	3
3/4	3/4	20	20	19	17	14	11
3/4	1	39	39	36	33	28	23
1	1	39	39	39	36	30	25
3/4	1 1/4	39	39	39	39	39	39
1	1 1/4	78	78	76	67	52	44
1 1/2	1 1/4	78	78	78	78	66	52
1	1 1/2	85	85	85	85	85	85
1 1/2	1 1/2	151	151	151	151	128	105
2	1 1/2	151	151	151	151	150	117
Over 60 PSI		40 ft.	60 ft.	80 ft.	100 ft.	150 ft.	200 ft.
3/4	1/2A	7	7	7	6	5	4
3/4	3/4	20	20	20	20	17	13
3/4	1	39	39	39	39	35	30
1	1	39	39	39	39	38	32
3/4	1 1/4	39	39	39	39	39	39
1	1 1/4	78	78	78	78	74	62
1 1/2	1 1/4	78	78	78	78	78	74
1	1 1/2	85	85	85	85	85	85
1 1/2	1 1/2	151	151	151	151	151	151
2	1 1/2	151	151	151	151	151	151

A: Min. building supply is 3/4 in.

NONPOTABLE WATER SUPPLY

Water conservation measures include the reuse of water collected from rainwater catchment systems, gray water drains, and recycled water for applications where drinking water standards are not required. Nonpotable water from these sources requires separate piping systems and other measures to ensure against contamination of the potable water supply. "Black water" from toilets, kitchen sinks, and dishwashers must always be directed to the sewer or on-site disposal system. Gray water can be obtained from other plumbing fixtures and be reused for landscape irrigation and for refilling toilet tanks.

General — 21 IRC / 21 UPC

- ☐ Permits required for nonpotable water systems ________ 2910.4 — 1501.3
- ☐ Operation & maintenance manual to be provided______2911.13 — 1501.6
- ☐ Diverter valves required for gray water systems ______ 2911.8.1 — 1503.2.2
- ☐ Gray water storage (surge) tanks require backwater valve (on piping from overflow to sewer)________________ 2911.8.2 — 1503.9.7
- ☐ Plot plan required for gray water systems ______________ local — 1503.5

System Identification

- ☐ Outlets (e.g., hose bibbs) require signage **F41**__ 2901.2.1 & 2910.3 — 601.3.5
- ☐ Distribution piping required to be purple_____________ 2901.2.2 — 601.3.3
- ☐ Piping requires label stating "CAUTION: NONPOTABLE WATER. DO NOT DRINK." (UPC: Fill in blank w/ type, such as GRAY WATER, RECLAIMED WATER, RAINWATER)_____________ 2901.2.2 — 601.3.3
- ☐ Label every 25 ft. (UPC: 20 ft.) & min. once each room 2901.2.2 — 601.3.2
- ☐ Lettering min. ½ in. high (¾-in. pipes 1½ in.–2 in.) __2901.2.2.2 — 601.3.2
- ☐ Labels to indicate direction of flow________________ 2901.2.2 — 601.3.2

FIG. 41 Nonpotable Water Supply Outlets

IRC Pictograph

Hose bibbs or other outlets of nonpotable water systems require a warning sign stating **"CAUTION: NONPOTABLE WATER. DO NOT DRINK"** *in letters min. ½ in. high and the accompanying pictograph.*

UPC Pictograph

PIPE INSULATION

Hot Water Piping: IRC — 21 IRC

- ☐ Hot water piping max. developed length 100 ft. from source of hot water to fixtures ____________________________ 2905.3[6]
- ☐ Heated water circulation systems require circulation pump_________1103.5.1.1
- ☐ Circulation system returns dedicated or to cold water pipe_________1103.5.1.1
- ☐ Returns to cold water pipe require controls to limit to 104°F_________1103.5.1.1
- ☐ Min. R-3 insulation required for the following: ___________________1103.5.2
  - Piping ≥¾ in. inside conditioned space
  - Piping serving ≥1 dwelling unit
  - Piping outside the conditioned space
  - Piping from the water heater to a distribution manifold
  - Buried piping or piping located under a floor slab
  - Supply & return piping in circulation systems (other than cold water return)

Hot Water Piping: UPC — 21 UPC

- ☐ Hot water piping requires insulation w/ thickness ≥ pipe diameter for pipes up to 2 in., 2 in. thick for larger pipes______ 609.12.1&2

6. New limitation on length of hot water pipe regardless of insulation.

CROSS-CONNECTION CONTROL

Backflow prevention devices protect water systems from backup and contamination. An air gap is a physical separation and is used to protect waste receptors, such as sinks. Atmospheric vacuum breakers (AVBs) **F45** prevent contaminants from entering through systems such as lawn sprinklers and must be in locations above the elevation of the potential contaminant. A reduced-pressure principle backflow protection assembly (RP) **F43** consists of two independently acting check valves, a differential pressure-relief valve, test cocks, and isolation valves. These afford the greatest degree of protection. Municipalities sometimes require them to protect the public system from contamination by a customer. They require testing upon installation and must be at least 12 inches above grade or floor.

Protection of Potable Water — 21 IRC / 21 UPC

- ☐ Prevent contamination of potable water supply ________2902.1 — 602.1
- ☐ Connections from private to public water supply prohibited2902.1 — 602.2
- ☐ Fixture outlet receptor air gaps:
  - Min. 2× diameter of outlet and per table **F42,T30** __ 2902.3.1 — 603.3.1

FIG. 42 Air Gap

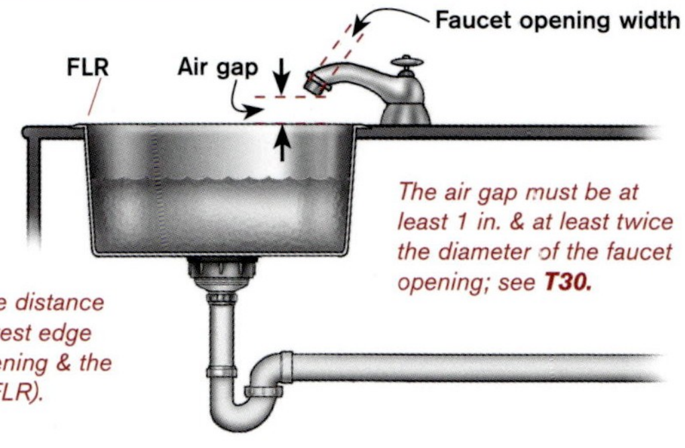

The air gap must be at least 1 in. & at least twice the diameter of the faucet opening; see ***T30.***

The air gap is the distance between the lowest edge of the faucet opening & the flood level rim (FLR).

TABLE 30 MINIMUM AIR GAPS ◆ IRC T2902.3.1 & UPC T603.3.1

Max. Opening Diameter & Typical Fixtures (in.)	Not Affected by Side Walls (in.)		Affected by Side Walls[A] (in.)	
	IRC	UPC	IRC	UPC
≤½ (lav)	1	1	1½	1½
≤¾ (LT)	1½	1½	2½	2¼
≤1(BT)	2	2	3	3
>1 (pool)	2× diameter	2× diameter	3× diameter	3× diameter

A. Affected by side walls = any time the distance from the spout to the wall is <3× the diameter of the effective opening, or <4× the diameter for 2 intersecting walls.

FIG. 43 Reduced Pressure Principle Backflow Prevention Assembly

RP device used for:
- *Boilers*
- *Fire-sprinkler systems*
- *Lawn irrigation systems*
- *Solar heating piping*
- *Isolation of private from public system*

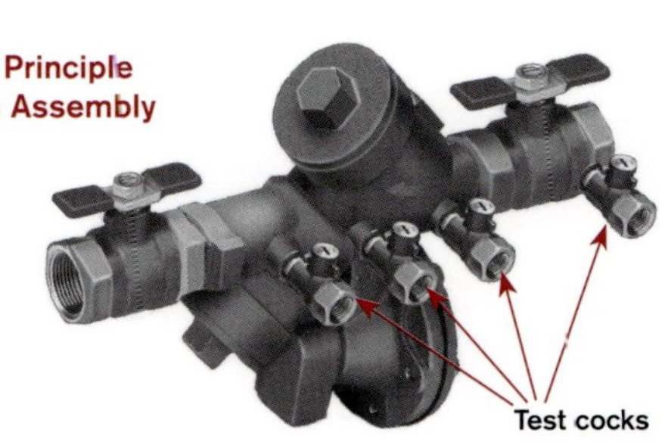

FIG. 44

Hose Bibb Backflow Protection

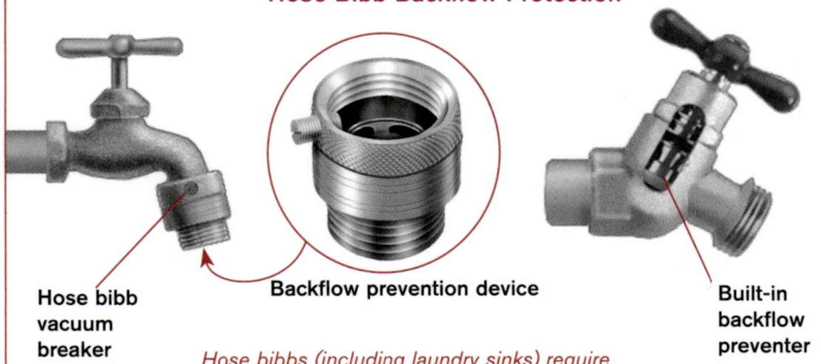

Hose bibbs (including laundry sinks) require backflow protection devices, except for clothes washers & tank drains (such as water heaters).

Backflow Prevention Devices

	21 IRC	21 UPC
☐ RP required **F43** for:		
• Boilers w/ conditioning chemicals	2902.5.1	603.5.10
• Fire-sprinkler systems w/ additives	2902.5.4.1	603.5.14.2
• Lawn irrigation systems w/ chemical injectors	2902.5.3	603.5.6.3
• Solar heating piping w/ additives	2902.5.5.3	603.5.6.3
• Pool/spa makeup water—RP or air gap required	2902.4	603.5.20
☐ AVB required for:		
• Hose bibbs except tank drains or CW valves **F44**	2902.4.3	603.5.7
• Pool/spa inlets min. 6 in. above critical level	2902.3.2	603.5.5
• Irrigation system, 6 in. above highest head	2902.3.2	603.5.6
• Pressure-type AVBs min. 12 in. above highest head **F45**	2902.3.4	603.5.6
• Flushometer valves, 6 in. above fixture FLR	2902.3.2	603.5.1
☐ Integral air gaps in fixtures to recognized standards OK for:		
• Reverse osmosis drinking water treatment units	2909.2	603.4.6
• DWs (UPC: comply w/ UL 749 section SA 3.4)	2717.1	414.2
(Air gaps on the supply are not the same as drainage air-gap devices **F67***).*		
• Pullout spouts and sprayers w/ integral AVB AMI	2902.4.2	603.4.6
• Pullout or separate shower spray wands	2902.4.2	603.4.6
• Flush tank w/ antisiphon fill valves (ballcock valves) critical level min. 1 in. above opening of overflow pipe	2902.4.1	603.5.2

FIG. 45

Atmospheric Vacuum Breaker

Pressure-type AVB is depicted. Nonpressure type would require 6 in. above head & no valve downstream of AVB.

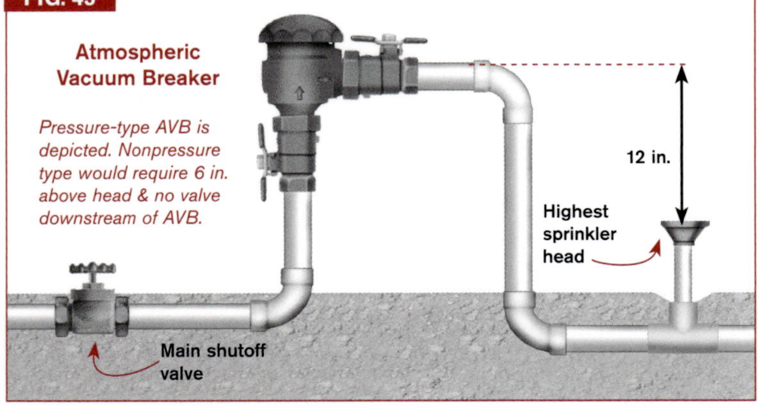

WATER PRESSURE

Excessive water pressure is controlled by installation of a pressure-reducing regulator on the incoming line. These regulators must conform to the ASSE 1003 standard, which includes a requirement for a screen that may require periodic cleaning. A "closed system" is created when a one-way valve is installed on the main water supply. Filters, backflow preventers, and regulators have such valves.

As the temperature recovers in a hot water storage tank, pressure increases and is normally equalized with the municipal water system pressure. Closed systems prevent this pressure equalization. The solution is an expansion tank installed on the cold water line at the water heater.

FIG. 46

Pressure Regulator

Pressure-reducing valves must conform to ASSE 1003 or CSA B356.

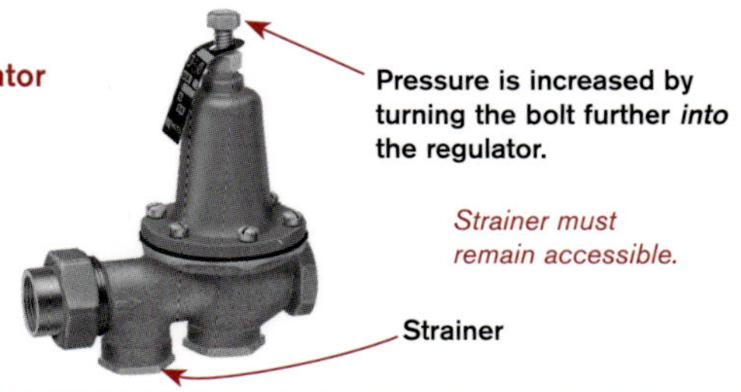

Strainer must remain accessible.

General

	21 IRC	21 UPC
☐ Public water main or individual water supply pressure must provide required flow at fixtures **T24**	2903.3	608.1
☐ If pressure insufficient, install booster pump, tanks, etc.	2903.3	608.1
☐ UPC min. 15 psi after allowing for all pressure losses	n/a	608.1
☐ Max. building water pressure 80 psi	2903.3.2	608.2
☐ PRR **F46** required when > 80 psi	2903.3.2	608.2
☐ Strainer **F46** required ahead of regulator EXC	2903.3.2	608.2
• Regulators for piping >1½ in.	n/a	608.2
☐ Regulator & strainer accessible w/o removing piping	MFR	608.2
☐ Pipe sizing based on 80% of regulated pressure	n/a	608.2
☐ Device for controlling pressure (UPC: Expansion tank) required on systems w/ regulators that prevent pressure dissipation **F47**	2903.4.1	608.3
Note: Older regulators with an "integral bypass feature" are no longer recognized by codes & should also be considered as creating a closed system.		
☐ Device for controlling pressure (UPC: Expansion tank) required if backflow device prevents pressure dissipation	2903.4.2	608.3

FIG. 47

Expansion Tank

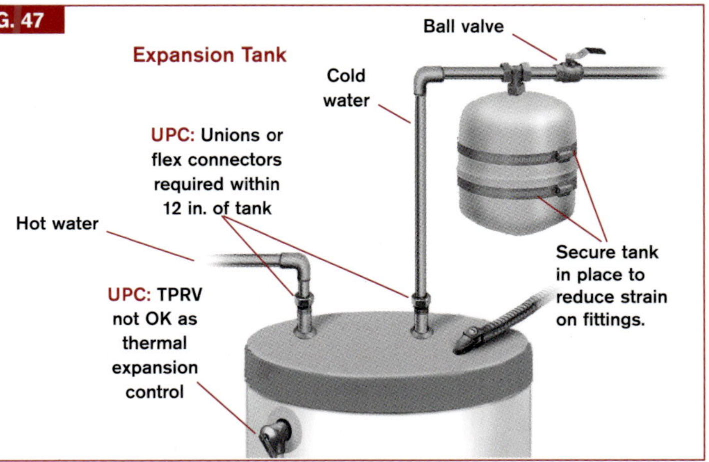

GAS PIPING

The model code upon which others are based is the *NFPA 54, The National Fuel Gas Code, ANSI Z223.1*. There are few differences between it and the fuel gas provisions of the IRC, IFGC, UPC, and UMC. This conformity allows appliance manufacturers and installers to have uniform nationwide standards and practices.

General — 21 IRC / 21 UPC

- ☐ Nonsteel pipe requires yellow label marked "gas" in black letters at ≤5-ft. intervals EXC ______ 2412.5 — n/a
  - When located in same room as appliance served ______ 2412.5 — n/a
- ☐ If >1 meter, mark each meter as to the premise served ______ 2412.7 — 1208.7.4
- ☐ LPG storage & piping per NFPA 58 (***p. 28***) ______ 2412.2 — 1212.11
- ☐ No piping in circulating air duct, chimney or gas vent, ventilating duct, or elevator shaft ______ 2415.3 — 1210.3.4
- ☐ No concealed piping in solid partitions (concrete or CMU walls) except in chase ______ 2415.4 — 1210.4.2

Metallic Pipe Joints & Fittings

- ☐ ≥Schedule 40 threaded, flanged, brazed, or press-connected fittings, <schedule 40 press-connect, flanges, brazing, or welding 2414.9.1 — 1208.6.10.1
- ☐ Clear fittings of burrs; brush & blow out chips & scales ______ 2414.6 — 1208.6.8
- ☐ Concealed locations (see glossary) require threaded elbows, tees & couplings, brazed or welded fittings, or fittings listed for the purpose: no unions, bushings, or compression couplings ______ 2415.5 — 1210.4.1

Materials

- ☐ Steel & wrought iron min. schedule 10 (UPC 40) ______ 2414.3.2[7] — 1208.6.3.1
- ☐ Cu tubing type K or L ______ 2414.4.3 — 1208.6.4.3
- ☐ No Cu or brass if >0.3 grains H_2S per 100 cu. ft. gas 2414.4.3 — 1208.6.4.3
- ☐ No pipe repair—pipe w/ defects must be replaced ______ 2414.6 — 1208.6.8
- ☐ Plastic pipe only OK outdoors & underground EXC ______ 2415.17.1 — 1210.1.7
  - Anodeless risers or wall head adapters ______ 2415.17.1X — 1210.1.7X
- ☐ Provide yellow insulated tracer wire min. #18 (UPC: #14) along plastic pipe & terminating above ground ______ 2415.17.3 — 1210.1.7.2
- ☐ CSST per approvals, listing & AMI ______ 2415.2 — 1208.6.4.5

Piping Support

- ☐ Max. support intervals for gas pipe & smooth tubing **T31** 2424.1 — 1210.3.5.1
- ☐ CSST support AMI ______ 2424.1 — 1210.3.5.1
- ☐ Hangers to prevent stress & dampen excessive vibration 2418.2 — 1210.3.5
- ☐ Hangers must allow for pipe expansion & contraction ______ 2418.2 — 1210.3.5.2
- ☐ Piping outdoors or on roofs elevated (IRC: min. 3½ in.) ______ 2415.9 — 1210.3.5.3

TABLE 31 — GAS PIPING & TUBING SUPPORT[A] IRC T2424.1 & UPC T1210.3.5.1

Steel Pipe Nominal Size (in.)	Max. Support Spacing (ft.)	Smooth-Wall Tubing Nominal Size (in.)	Max. Support Spacing (ft.)
½	6	½	4
¾ or 1	8	⅝ or ¾	6
≥1¼ (horizontal)	10	⅞ or 1 (horizontal)	8
≥1¼ (vertical)	every floor level	⅞ or 1 (vertical)	every floor level

A. CSST support AMI, not to this table.

7. Previous code editions limited the size to schedule 40 for steel pipe & black pipe. That would still be the case for threaded pipe. Schedule 10 can be connected w/ press fittings.

Protection & Installation — 21 IRC / 21 UPC

- ☐ Outdoor piping min. 3½ in. above ground or roof surface 2415.9 — n/a
- ☐ Piping in light-frame construction requires shield plates to protect against fastener penetration if <1½ from face of framing EXC 2415.7.1&2 — 1210.4.3
  - Black or galvanized steel piping ______ 2415.7X — 1210.4.3
- ☐ Shield plates min. 16 ga or per MFR ______ 2415.7.3 — 1210.4.3
- ☐ Extend shield plates 4 in. horizontally past pipe/tubing ______ 2415.7.1 — 1210.4.3
- ☐ Extend plates 4 in. vertically above bottom plate & 4 in. down from top plate ______ 2415.7.1 — n/a

FIG. 48 Shield Plates

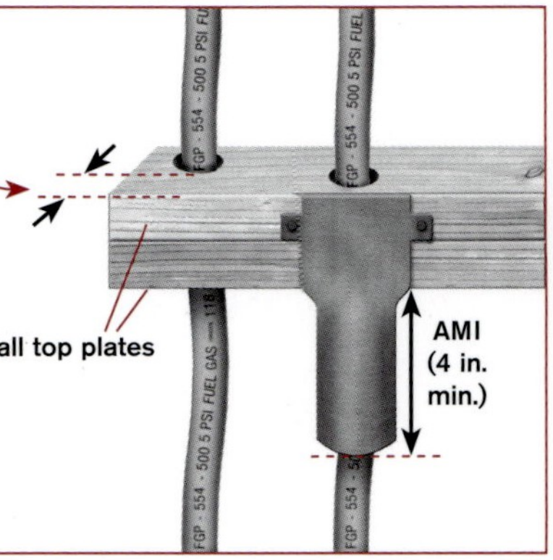

Piping & tubing other than black or galvanized pipe requires shield-plate protection. Tubing through bored holes can also be protected by a galvanized pipe sleeve or conduit secured in place.

Testing — 21 IRC / 21 UPC

- ☐ All piping installations to be pressure tested EXC ______ 2417.1 — 1213.1
  - Minor repairs & additions & new branches can be tested w/ non-corrosive leak-detecting fluid or other methods 2417.1.2&3 — 1213.1.2
- ☐ Test medium air, nitrogen, or CO_2 (not oxygen) ______ 2417.2 — 1213.3
- ☐ Leave all joints exposed until tested ______ 2417.3 — 1213.2.1
- ☐ IRC: Test min. 1½ × working pressure, min. 3 psig ______ 2417.4.1 — n/a
- ☐ IRC: Test duration min. 10 minutes ______ 2417.4.2 — n/a
- ☐ UPC: Test pressure min. 10 psig (MP CSST: Min. 30 psig) ______ n/a — 1213.3
- ☐ UPC: Test time min. 15 minutes (MP CSST: Min. 30 min.) ______ n/a — 1213.3
- ☐ Range of gauge scale ≤5× (UPC: ≤ 2×) test pressure ______ 2417.4 — 318.5
- ☐ If existing appliances not rated for test pressure, isolate by disconnecting & capping their outlets ______ 2417.3.3 — 1213.2.4
- ☐ If existing appliances rated for test pressure, isolation by closing their valves OK ______ 2417.3.4 — 1213.2.5
- ☐ Use approved non-corrosive leak-detecting fluid or gas detector for locating leaks ______ 2417.5.1 — 1213.4.1
- ☐ Matches, open flames not OK for locating leaks ______ 2417.5.1 — 1213.4.1
- ☐ Inspect for open fittings or valves before turning on gas 2417.6.2 — 1213.5.1
- ☐ Check for leakage immediately after turning on gas ______ 2417.6.3 — 1213.5.2

Electrical Bonding other than CSST

- ☐ Electrical bond required for above-ground gas piping ______ 3609.7 — 1211.1
- ☐ EGC of circuit supplying equipment OK as bond ______ 3609.7 — 1211.1
- ☐ Gas piping not OK as grounding electrode in earth ______ 2410.1 — 1211.4

Gas Piping Underground

	21 IRC	21 UPC
☐ Protect metallic piping from corrosion	2415.11	1210.1.3
☐ Zi coatings not sufficient as corrosion protection	2415.11.1	1210.1.3.1
☐ Factory-applied electrically insulating coatings AMI	2415.11.2	1210.1.3.2
• Field application of nipples & fittings OK if AMI	2415.11.2	1210.1.3.2
(Note: The AHJ may require observation of field wrapping after pressure testing.)		
☐ Cathodic protection for steel risers from plastic pipe EXC	2415.11.4	1210.1.3.9
• Anodeless risers	2415.5.1 & 2415.11.4	1210.1.3.9
☐ Monitor & maintain cathodic protection	2415.11.2	1210.1.3.3
☐ Min. cover depth 12 in. (UPC: 18 in. if damage likely)	2415.12	1210.1.1
☐ If <12 in., provide conduit or bridge	n/a	1210.1.1
☐ 8-in. cover OK for single appliance if damage not likely	2415.12.1	n/a
☐ Pipe trenches to have firm continuous bearing	2415.13	1210.1.2
☐ Piping not to penetrate foundation below grade (UPC: OK if sleeved & sealed)	2415.6	1210.1.5

Underground beneath Buildings

	21 IRC	21 UPC
☐ Piping in slab requires protected channel or conduit	2415.8	n/a
☐ Piping underground beneath buildings requires protective conduit **F49**	2415.14	1210.1.6
☐ Conduit w/ one end on exterior & one on interior: Pipe sealed gastight to conduit at interior, exterior pipe terminated min. 4 in. outside building & conduit vented above grade and configured to prevent water & insect entry **F49**	2415.8.1 & 2415.14.1	1210.1.6.1
☐ Conduit w/ both ends terminating inside building should not have ends sealed & both must be accessible	2415.8.2 & 2415.14.2	1210.1.6.2

FIG. 49 Gas Piping below Building Slab

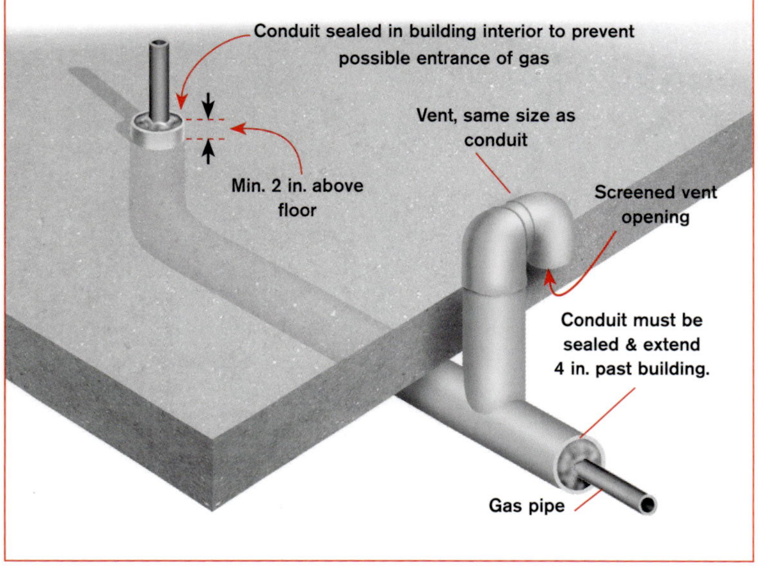

Drips & Sediment Traps

	21 IRC	21 UPC
☐ Slope piping min. 1/4 in./15 ft. unless dry gas	2419.1	1210.3.3
☐ If non-dry gas, install drips to prevent moisture from reaching meter	2419.2	1210.7
☐ Sediment traps required as close as practical to appliance inlets **F50** EXC	2419.4	1212.9[8]
• Ranges, dryers, gas lights, fireplaces & outdoor grills	2419.4	1212.9

8. Previous edition also required that the sediment trap be upstream of any flex connector.

Valves

	21 IRC	21 UPC
☐ All valves accessible & protected from damage	2420.1.3	1212.6
☐ Valve within 6 ft. of appliance & in same room EXC	2420.5.1	1212.6
• Valve at manifold within 50 ft. & permanently marked	2420.5.3	Ø
☐ Valve upstream from union or connector at appliance	2420.5.1	1212.6
☐ Shutoff valves in tubing systems rigidly supported independently of the tubing	2420.6	MFR
☐ Shutoff valves required outdoors at each building	2420.3	1210.9.2
☐ Each appliance requires its own shutoff valve	2420.5	1212.6
☐ Valve required ahead of gas connector	2422.1.2.4	1212.6
☐ Listed valve required ahead of each MP regulator **F51**	2420.4	1210.9
☐ Valve for decorative appliances in fireplace install AMI (UPC: must be listed for such use)	2420.5.1	1212.6X1
☐ Valve in fireplace must be removed & pipe capped if fireplace to be used with solid fuels	n/a	1210.8.2
☐ Cap any unused outlets (valve alone not OK)	2417.6.2	1210.8.1

Appliance Connections

	21 IRC	21 UPC
☐ Connectors rigid pipe, L&L flexible connectors, CSST AMI to appliance fixed in place, or listed outdoor gas hose connectors	2422.1	1212.1
☐ Connector max. 6-ft. length EXC	2422.1.2.1	1212.6
• Rigid piping w/ valve ≤6 ft. of appliance served & sized as piping rather than as connector	2422.1.2.1X	n/a
☐ Flexible connectors max. one per appliance	2422.1.2.1	1212.1(3)
☐ Flex connectors in same room as appliance	2422.1(3)	1212.1(3)
☐ Flex connectors not to pass through walls, walls, partitions, ceilings, or floors EXC	2422.1.2.3	1212.5.3
• Semi-rigid tubing & listed connectors OK to pass through opening in appliance housing if protected against damage	2422.1.2.3X4	1212.1(3)
☐ Connector's capacity ≥ appliance demand **T32**	2422.1.2.2	1212.1(3)
☐ Flex connectors one-time use only—do not reuse	2422.1(3)	1212.1(3)
☐ Flex connectors rated for max. 0.5 psig or AMI	2422.1.2	1212.1(3)

TABLE 32 TYPICAL[A] FLEXIBLE GAS CONNECTOR CAPACITY

Nom. O.D.[B]	Nom. I.D.[B]	Lengths (in.) & Capacity (Btu/hr.)							
		12	18	24	30	36	48	60	72
3/8	1/4	48,000	43,800	40,000	36,400	33,400	28,300	24,900	23,100
1/2	3/8	102,000	93,100	85,000	77,100	71,100	60,500	53,200	49,100
5/8	1/2	180,000	164,200	150,000	136,000	125,000	106,000	93,200	86,000
7/8	3/4	290,900	290,900	290,900	270,500	255,900	215,000	197,400	173,900

A. Based on typical MFR instructions for standard (not high-capacity) connectors, gas at 1K/Btu/cu. ft., gas at 0.5 in water column pressure drop.
B. O.D. = outside diameter in., I.D. = inside diameter in.

FIG. 50

Sediment Trap

Place sediment traps as close as practical to appliance inlets. The sediment trap must cause a 90° turn in the direction of the gas.

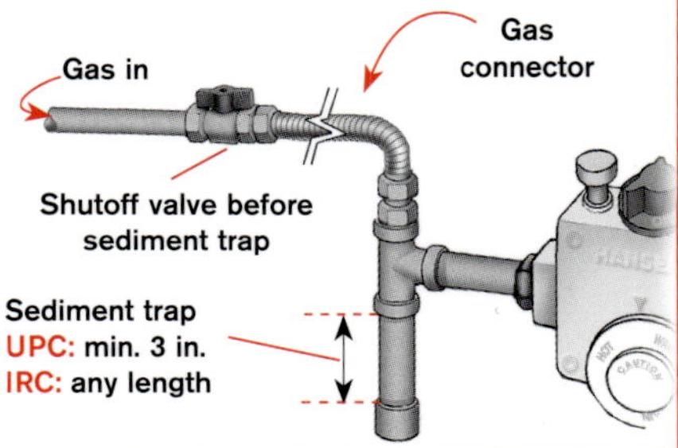

Corrugated Stainless-Steel Tubing: CSST — 21 IRC — 21 UPC

- ☐ Follow MFR instructions—install AMI: ______ 2415.2 — 1210.2
  - Size, support & bend radius AMI per MFR tables
  - Route below grade using conduit unless specifically allowed AMI
  - Proprietary shield plates per MFR **F48**
  - Avoid kinking, twisting, or contact w/ sharp objects

Electrical Bonding for CSST

- ☐ Arc-resistant CSST must be electrically continuous & connected to effective ground-fault current path ______ 2411.3 — 1211.3[9]
- ☐ Arc-resistant CSST considered bonded when connected to appliances w/ equipment grounding conductor ______ 2411.3 — 1211.3[9]
- ☐ If any CSST component lacks arc-resistant jacket or coating, bonding is required same as for other CSST ______ 2411.3 — 1211.3[9]
- ☐ Piping systems containing 1 or more segments of CSST require bond to electrical service grounding electrode system or to lightning protection system ______ 2411.2 — 1211.2
- ☐ Connect bonding jumper to metal pipe, metal pipe fitting, or CSST fitting ______ 2411.2.1 — 1211.2.1
- ☐ Bonding jumper min. #6 Cu wire ______ 2411.2.2 — 1211.2.2
- ☐ Bonding jumper max. 75 ft. length ______ 2411.2.3 — 1211.2.3
- ☐ Bonding jumper clamps listed for the purpose ______ 2411.2.5 — 1211.2.4&5

Medium Pressure (MP) Regulators F51

- ☐ Regulators required when line pressure exceeds allowed inlet pressure of appliances served ______ 2421.1 — 1208.8
- ☐ Regulators require listing to ANSI Z21.80 ______ 2421.1 — 1208.8.1
- ☐ MP regulators must be accessible ______ 2421.1&2 — 1208.8.2
- ☐ Capped or plugged tee fitting required between regulator & its upstream shutoff valve ______ 2421.2 — MFR
- ☐ Capped tee fitting required downstream of regulator ______ 2421.2 — MFR
- ☐ Vented regulators must be vented to outdoors EXC ______ 2421.3 — 1208.8.4(1)
  - If equipped w/ approved vent-limiting device ______ 2421.3X — 1208.8.4(1)X

FIG. 51

CSST Manifold

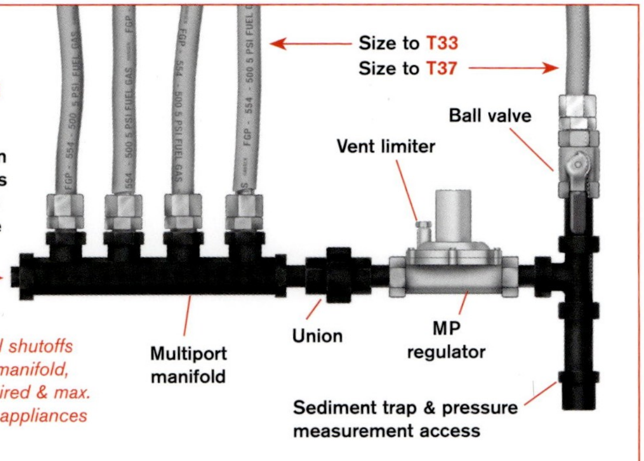

The standard for CSST, CSA LC-1, requires that workers be certified to install CSST. The manufacturers offer training & certification courses. Installers should carry their certification card and present it to inspectors upon request.

FIG. 52

Gas Meters and CSST Bonding

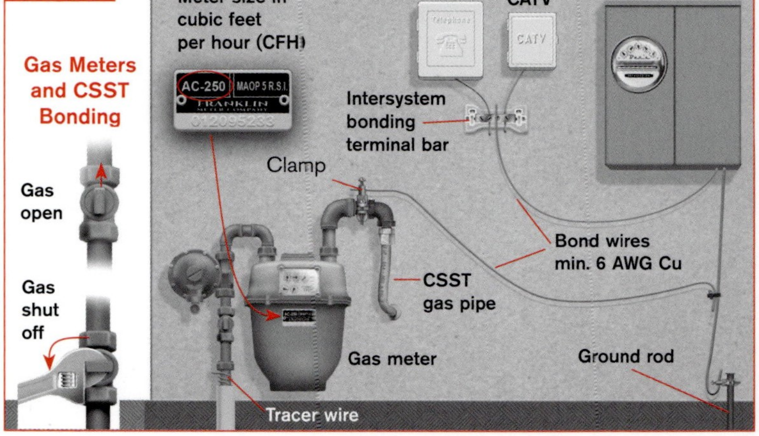

The NEC only requires gas pipe bonding to ensure an effective fault-current path, not as a path for lightning. Concern over damage from indirect lightning strikes has resulted in requirements for bonding beyond the NEC requirements. The bonding connection for tubing that is not listed with an arc-resistant jacket or coating must be to a metallic pipe, pipe fitting, or CSST fitting. The bonding conductor can connect to the electrical service grounding electrode system, to a tap on the grounding electrode conductor (as shown here), or to a lightning protection grounding electrode system. It should not connect to the intersystem bonding terminal bar (IBT), which is reserved for systems such as telephone and cable TV.

TABLE 33 — CUBIC FEET CAPACITY OF CSST[A] – LOW PRESSURE IRC T2413.4(5) & UPC T1215.2(14)

Pipe Length (ft.)	Effective Hydraulic Diameter[B] (EHD) and Corresponding Approximate Nominal Tubing Size[C]									
	13	18	19	23	25	31	37	39	46	62
	3/8 in.	1/2 in.	1/2 in.	3/4 in.	3/4 in.	1 in.	1 1/4 in.	1 1/4 in.	1 1/2 in.	2 in.
	Demand Capacity (in cu.ft./hr.)									
5	46	115	134	225	270	546	895	1037	1790	4,140
10	32	82	95	161	192	383	639	746	1260	2,530
15	25	66	77	132	157	310	524	615	1030	2,400
20	22	58	67	116	137	269	456	536	888	2,080
25	19	52	60	104	122	240	409	482	793	1,860
30	18	47	55	96	112	218	374	442	723	1,700
40	15	41	47	83	97	188	325	386	625	1,470
50	13	37	42	75	87	168	292	347	559	1,320
60	12	34	38	68	80	153	267	318	509	1,200
70	11	31	36	63	74	141	248	295	471	1,110
80	10	29	33	60	69	132	232	277	440	1,040
90	10	28	32	57	65	125	219	262	415	983
100	9	26	30	54	62	118	208	249	393	933
150	7	20	23	42	48	91	171	205	320	762
200	6	18	21	38	44	82	148	179	277	661
250	5	16	19	34	39	74	133	161	247	591
300	5	15	17	32	36	67	95	148	226	540

A. Based on inlet pressure < 2 psi, pressure drop 0.5-in. water column, specific gravity 0.60.
B. EHD is a measure of relative efficiency. These values vary from one manufacturer to another. The tables are given here to determine approximate tubing sizes. These tables include 4 90° bends and 2 end fittings. Additional bends require adjustment in accordance w/ the manufacturer's instructions.
C. EHD values for tubing sizes vary from one MFR to another, hence there are multiple columns for identical tubing sizes. If MFR values do not align w/ this table, follow the tables provided by the MFR.

9. Previous code edition recognized CSST installed AMI in accord w/ its listing. This section provides additional clarity regarding arc-resistant jacketed & coated systems.

GAS PIPE SIZE

Gas pipe sizing is a function of aggregate appliance demand, gas pressure, BTUs per cubic foot of gas, and the pipe run length. Corrugated stainless-steel tubing systems (CSST) can be run at medium pressure (approximately 2 psig) to a central manifold **F51** where the pressure is reduced to the operating pressure of appliances, typically not more than 0.5 psig. This method allows smaller main runs and uses less tubing than a system where all piping is operated at utilization pressure.

Gas piping systems in series **F53** can be sized using either the longest length or the branch length method. Systems with MP regulators are sized using the hybrid pressure method. There are separate tables for steel pipe and CSST tubing. The corrugations of CSST reduce its effective interior diameter, and its size is converted to "equivalent hydraulic diameter." In general, it must be 1 size larger than the corresponding size of steel pipe to have the same capacity.

Pipe Size	**21 IRC**	**21 UPC**
☐ Size per max. demand based on appliance input ratings	2413.2	1208.4.1
☐ Assume all appliances operating simultaneously EXC	2413.2	1208.4.1
• Where diversity of load can be established	2413.2	1208.4.1X
☐ Size AMI or per tables **T34,35,36,38**	2413.3	1208.4.2
☐ Adjust volumetric flow rate for altitudes > 2,000 ft.	2413.2	1208.4.1

TABLE 34 GAS PIPE SIZING PROCEDURE ◆ IRC 2413.3 & UPC 1215.1

1. Determine Btu/cu.ft. from local gas provider.
2. Determine cu.ft./hr. demand for each appliance.
3. Sketch layout w/ piping lengths to each appliance **F53**.
4. Determine total cu.ft./hr. demand on each pipe section.
5. Determine length to most remote appliance.
6A. (longest length method) Use row of **T36** for that length for all appliances.
6B. (branch length method) Use same row for all sections in series w/ most remote appliance. For other branches, use actual length of each branch.

The "longest length" method is more conservative and compensates for pressure losses throughout the system. The "branch length method" has less leeway, and consideration should be given to the lengths of pipe fittings. The codes accept both methods. Systems w/ MP regulators use the "hybrid pressure" method, where the pipe sizes before the regulator are determined separately, each by the longest length method.

*To use the fill-in table below, first create a diagram such as the example in **F53**. If the Btu per cubic foot capacity of the gas is not known, use a 1:1 ratio, which also simplifies the equations and in most cases yields a more conservative result. Next, assign the cubic foot values for each appliance and the cubic foot for each section of piping. Working from the most remote appliance back toward the meter, assign pipe sizes to each section of the diagram.*

TABLE 35 GAS PIPE SIZE EXAMPLE FILL-IN

Pipe Section	Total cu. ft./hr.[A]	Longest Length	Longest Length Method	Actual Length	Branch Length Method
A					
B					
C					
D					
E					
F					

A. Btu/cu.ft. (from gas supplier).

TABLE 36 CUBIC FEET CAPACITY OF SCHEDULE 40 METALLIC GAS PIPE[A] ◆ IRC T2413.4(1) & UPC T1215.2(1)

Pipe Length (in ft.)	Nominal Pipe Size (in.)								
	½	¾	1	1¼	1½	2	2½	3	4
	Demand Capacity (in cu.ft./hr.)								
10	172	360	678	1,390	2,090	4,020	6,400	11,300	23,100
20	118	247	466	957	1,430	2,760	4,400	7,780	15,900
30	95	199	374	768	1,150	2,220	3,530	6,250	12,700
40	81	170	320	657	985	1,900	3,020	5,350	10,900
50	72	151	284	583	873	1,680	2,680	4,740	9,660
60	65	137	257	528	791	1,520	2,430	4,290	8,760
70	60	126	237	486	728	1,400	2,230	3,950	8,050
80	56	117	220	452	677	1,300	2,080	3,670	7,490
90	52	110	207	424	635	1,220	1,950	3,450	7,030
100	50	104	195	400	600	1,160	1,840	3,260	6,640
125	44	92	173	355	532	1,020	1,630	2,890	5,890
150	40	83	157	322	482	928	1,480	2,610	5,330
175	37	77	144	296	443	854	1,360	2,410	4,910
200	34	71	134	275	412	794	1,270	2,240	4,560
250	30	63	119	244	366	704	1,120	1,980	4,050
300	27	57	108	221	331	638	1,020	1,800	3,670
350	25	53	99	203	305	587	935	1,650	3,370
400	23	49	92	189	283	546	870	1,540	3,140
450	22	46	86	177	266	512	816	1,440	2,940
500	21	43	82	168	251	484	771	1,360	2,780

A. Based on inlet pressure <2 psi, pressure drop 0.5-in. water column, specific gravity 0.60.

TABLE 37 CUBIC FEET CAPACITY OF CSST[A] – MEDIUM PRESSURE ◆ IRC T2413.4(6) & UPC T1215.2(17)

Pipe Length (ft.)	Effective Hydraulic Diameter[B] (EHD) and Corresponding Approximate Nominal Tubing Size[C]									
	15	18	19	23	25	31	37	39	46	62
	⅜ in.	½ in.	½ in.	¾ in.	¾ in.	1 in.	1¼ in.	1¼ in.	1½ in.	2 in.
	Demand Capacity (in cu.ft./hr.)									
10	353	587	700	1,100	1,370	2,990	4,510	5,037	9,600	21,600
25	220	374	444	709	876	1,870	2,890	3,258	6,040	13,700
30	200	342	405	650	801	1,700	2,640	2,987	5,510	12,500
40	172	297	351	567	696	1,470	2,300	2,605	4,760	10,900
50	154	266	314	510	624	1,310	2,060	2,343	4,260	9,720
75	124	218	257	420	512	1,070	1,690	1,932	3,470	7,940
80	120	211	249	407	496	1,030	1,640	1,874	3,360	7,690
100	107	189	222	366	445	920	1,470	1,685	3,000	6,880
150	87	155	182	302	364	748	1,210	1,389	2,440	5,620
200	75	135	157	263	317	645	1,050	1,212	2,110	4,870
250	67	121	141	236	284	576	941	1,090	1,890	4,360
300	61	110	129	217	260	525	862	999	1,720	3,980

A. Based on inlet pressure 2.0 psi, pressure drop 0.5-in. water column, specific gravity 0.60. This table does not include effect of pressure drop across the line regulator. Where regulator losses exceed ¾ psi, do not use this table; refer to MFR.

B. EHD is a measure of relative efficiency. These values vary from one manufacturer to another. The tables are given here to determine approximate tubing sizes. These tables include 4 90° bends and 2 end fittings. Additional bends require adjustment in accordance w/ the manufacturer's instructions.

C. EHD values for tubing sizes vary from one MFR to another, hence there are multiple columns for identical tubing sizes. If MFR values do not align w/ this table, follow the tables provided by the MFR.

FIG. 53 Gas Pipe Size Example

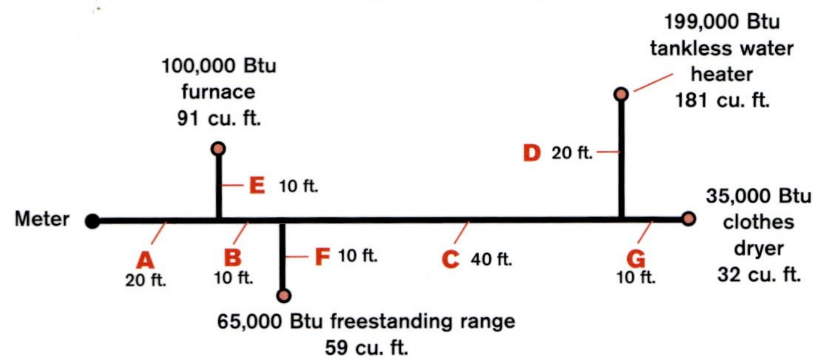

TABLE 38 GAS PIPE SIZE EXAMPLE

Pipe Section	Total cu. ft./hr.[A]	Longest Length	Longest Length Method	Actual Length	Branch Length Method
A	363	90 ft.	1¼ in.	90 ft.	1¼ in.
B	272	90 ft.	1¼ in.	90 ft.	1¼ in.
C	213	90 ft.	1¼ in.	90 ft.	1¼ in.
D	181	90 ft.	1 in.	90 ft.	1 in.
E	91	90 ft.	¾ in.	30 ft.	½ in.
F	59	90 ft.	¾ in.	40 ft.	½ in.
G	32	90 ft.	½ in.	80 ft.	½ in.

A. Example based on 1,100 Btu/cu. ft.—contact local provider for actual values.

PROPANE (LP GAS)

NFPA 58, the Liquefied Petroleum Gas Code, is the reference standard for LP gas systems. LP liquefies under moderate pressure and vaporizes upon release of the pressure. Horizontal storage tanks are manufactured to standards from the American Society of Mechanical Engineers (ASME), and portable cylinders are manufactured to U.S. Department of Transportation (DOT) standards. As a liquid or gas, propane is heavier than air, and when gas leakage occurs in a pit or basement, an invisible pool of combustible material can accumulate until it rises to the level of an ignition source, such as the pilot or igniter on an appliance. One method of protection against this hazard is an interlock that would shut off the gas flow. Another method is to install a pan drain below the equipment with a duct sloped to the exterior.

Horizontal ASME Tanks — NFPA 58

- ☐ Tank clearances **F54** ____ T6.4.1.1 T6.4.4.3 & Annex I
- ☐ Tank not allowed indoors ____ 6.2.1
- ☐ Protect tanks from damage (bollards) where ≤10 ft. of vehicle path ____ 6.8.1.2
- ☐ Masonry or concrete foundation required under tanks ____ 6.8.3.1
- ☐ Supports & saddles must be corrosion resistant & noncombustible ____ 6.8.3.4
- ☐ Secure tank against flotation in flood hazard areas ____ 6.8.1.6
- ☐ Secure tanks in seismic areas per ASCE SEI 7 as approved by AHJ ____ 5.2.4.4D
- ☐ Min. 10 ft. from easily ignitable material (firewood) ____ 6.5.3.3
- ☐ Replace containers w/ excessive dents or corrosion ____ 5.2.1.4

Propane Piping & Tubing Systems — NFPA 58

- ☐ Material black pipe, galvanized, brass, PE, SS, Cu, or polyamide AMI ____ 5.11.3.1
- ☐ Tubing brass, Cu L or K, CSST, PE, or polyamide AMI ____ 5.11.3.2
- ☐ Plastic tubing and fittings only outdoors & underground ____ 6.11.4.1
- ☐ Min. #14 tracer wire required to be run w/ plastic pipe ____ 6.11.4.6
- ☐ No cast-iron pipe fittings ____ 5.11.4.2
- ☐ Buried metal pipe min. 12-in. cover (18 in. if damage likely) ____ 6.11.3.13
- ☐ Conduit or shielding OK when 12-in. cover not possible ____ 6.11.3.13B
- ☐ Underground metal piping requires corrosion protection ____ 6.11.3.15
- ☐ LP gas piping not OK as grounding electrode ____ 6.11.3.16
- ☐ Grounding & bonding for static electricity protection not required ____ A6.25.1.3
- ☐ Dielectric fitting required between above- & below-ground pipe ____ 6.11.3.17
- ☐ Dielectric fitting to be above ground and outdoors ____ 6.11.3.17

FIG. 54 Propane Tank Container Spacing

Clearances are for tanks that are filled on site.

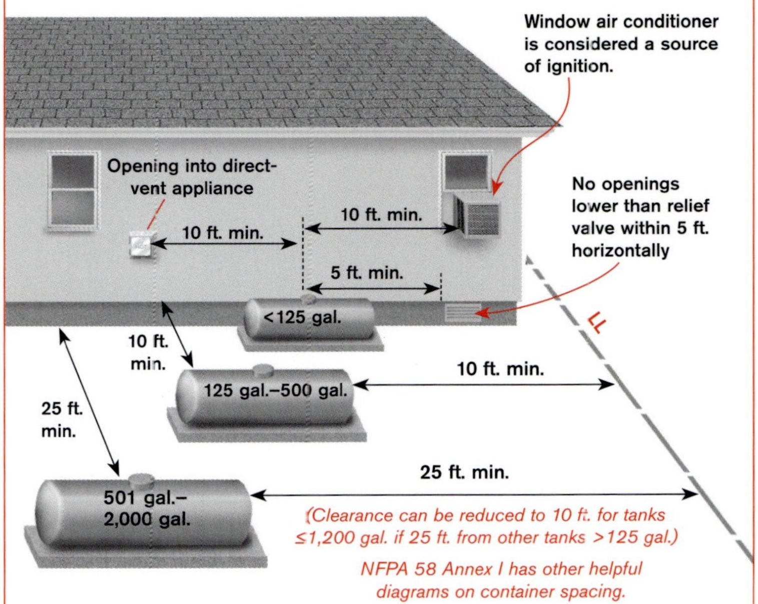

NFPA 58 Annex I has other helpful diagrams on container spacing.

Propane Tank Valves & Regulators — NFPA 58

- ☐ Vapor & liquid shutoff w/ internal excess flow valve required ____ T5.9.4.1
- ☐ Shutoff valve required to be readily accessible ____ 5.9.8.1G
- ☐ PRV required ____ 5.9.2.5 & 5.9.4.1
- ☐ Rain cap required over PRV ____ 6.9.2.4
- ☐ No shutoff valve between tank & PRV or on outlet of PRV ____ 6.9.2.7&9
- ☐ First-stage or high-pressure regulator must be installed outdoors ____ 6.10.1.3
- ☐ Regulator must be designed to resist elements (freezing) ____ 6.10.1.4

WATER HEATERS

Water heaters should be maintained at the lowest temperature needed to minimize scalding risk, but high enough to reduce the possibility of unwanted biologic growth in the system. An undersized water heater is more likely to be turned to a dangerously high setting. Protection against scalding can also be provided by tempering valves at the water heater or at individual fixtures. Tankless water heaters are becoming more popular, including hybrid systems that contain a small storage tank and circulating line. Heat pump water heaters may be encouraged or required by local energy codes and need a space with a large air volume for the extraction of heat from ambient air.

TABLE 39 — UPC WATER HEATER MIN. CAPACITY[A] ◆ T501.1(2)

Number of Bathrooms	Number of Bedrooms	1st-hr. Rating[B]
1 to 1½	1	38
	2-3	49
2 to 2½	2	49
	3-4	62
	5	74
3 to 3½	3	62
	4-6	74

A. Based upon the first-hour rating found on the "Energy Guide" label. This number is approximately equal to the storage size plus hourly recovery rate.
B. This table can also be used to size tankless water heaters.

General — 21 IRC / 21 UPC

- ☐ Replacement water heaters require permits ____________ 105.1 — 502.1
- ☐ Installation & maintenance instructions to be left w/ WH_ 1307.1 — 507.24
- ☐ Size to meet demand **T39** ________________________ 2448.1.1 — 501.1
- ☐ Installation AMI & all instructions in L&L _____ 2005.1 & 2801.3 — 501.1
- ☐ WH also used for space heating must be L&L for both __2448.2 — 501.1
- ☐ Electric WH requires in-sight or lockable disconnect **F61** T4101.5 — NEC
- ☐ Fuel-fired WH combustion air; see ***p. 41*** ____________ 2407.1 — 506.1
- ☐ Fuel-fired WH venting; see ***pp. 42–45*** ________ 1801.1 & 2427.1 — 509.0

Valves & Connections

- ☐ Full-open type valve required on supply at WH **F39,56** 2903.9.2 — 606.2
- ☐ Systems also used for space heating require temperature-actuated mixing valve to temper domestic water to 140°F or less _2803.2 — n/a
- ☐ Unions (UPC: Required within 12 in.) to allow removal **F57,61** MFR — 609.5
- ☐ Cu, brass, or stainless-steel flexible connectors max. 24 in. _ n/a — 604.12
- ☐ Where water heater tank located at elevation above fixture outlets (e.g., attic), vacuum-relief valve required on water heater___ n/a — 608.7
- ☐ PEX, PEX-AL-PEX, PE-AL-PE, or PE-RT tubing not within first 18 in. of piping connected to WH_________MFR — 604.13

FIG. 55

FVIR Water Heater

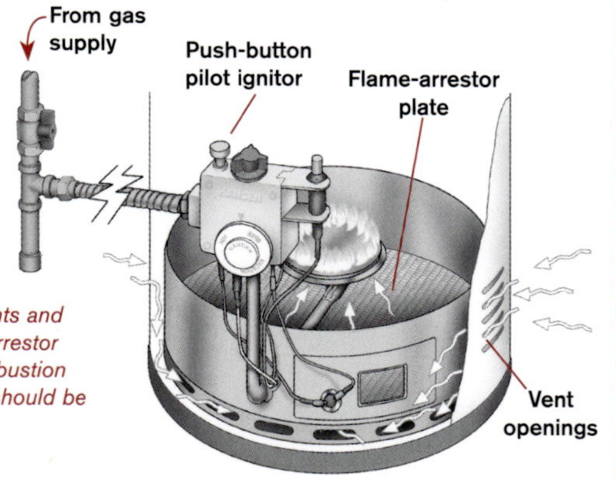

Air enters through the vents and passes through a flame-arrestor plate into the sealed combustion chamber. Vent openings should be cleaned periodically.

Special Locations — 21 IRC / 21 UPC

- ☐ Fuel-fired WH prohibited in storage closets __ 2005.2 & 2406.2 — local
- ☐ Not in bedrooms or bathrooms EXC ________ 2005.2 & 2406.2 — 504.1
  - Direct-vent WH OK w/o enclosure________ 2005.2 & 2406.2 — 504.1(1)
  - In dedicated enclosure w/ solid, weatherstripped, self-close door & all combustion air from exterior _____ 2005.2 & 2406.2 — 504.1(2)
- ☐ Ignition source ≥18 in. above garage floor EXC _______ 2801.7 — 507.13
  - Flammable Vapor Ignition-Resistant (FVIR) WH **F56** _2801.7X — 507.13
  - WH in separate enclosure accessible only from outside the garage & no combustion air from garage __ 1307.3 — 507.13.2
- ☐ Not to be installed where flammables are stored _________MFR — 507.12
- ☐ Seismic bracing required upper & lower ⅓ of tank in SDC D & townhomes SDC C (UPC: all occupancies SDC C–F) **F56** 1307.2 & 2801.8 — 507.2
- ☐ Barrier or elevation in vehicle path (garages, etc.) **F56** _ 1307.3.1 — 507.13.1
- ☐ Min. 3-in. concrete pad if supported on ground ____ 1305.1.3.1 — 507.4

Access & Working Space

- ☐ Clearances to combustibles per L&L and AMI 1306.1 & 2801.4 — 504.3.1
- ☐ Remain accessible for service, inspection & removal____1305.1 — 504.3.1
- ☐ Attic hatch or door min. 22 in. wide × 30 in. high___1305.1.2&3 — 508.4
- ☐ Largest appliance fits through access (crawl, attic)___1305.1.2&3 — 508.4
- ☐ Attic min. 24-in. passageway, solid floor to WH______ 1305.1.2 — 508.4.2
- ☐ Max. 20 ft. from access opening if ceiling <6 ft.______ 1305.1.2 — 508.4.1
- ☐ Min. 30-in. × 30-in. level working platform EXC ______ 1305.1.2 — 508.4.3
  - Platform not required if service possible from opening1305.1.2 — local
- ☐ Light & receptacle near WH in attic______________ 1305.1.2.1 — 508.4.4
- ☐ Light switch required at passageway entrance _________ 3903.4 — 508.4.4

FIG. 56

FVIR Water Heater in Garage

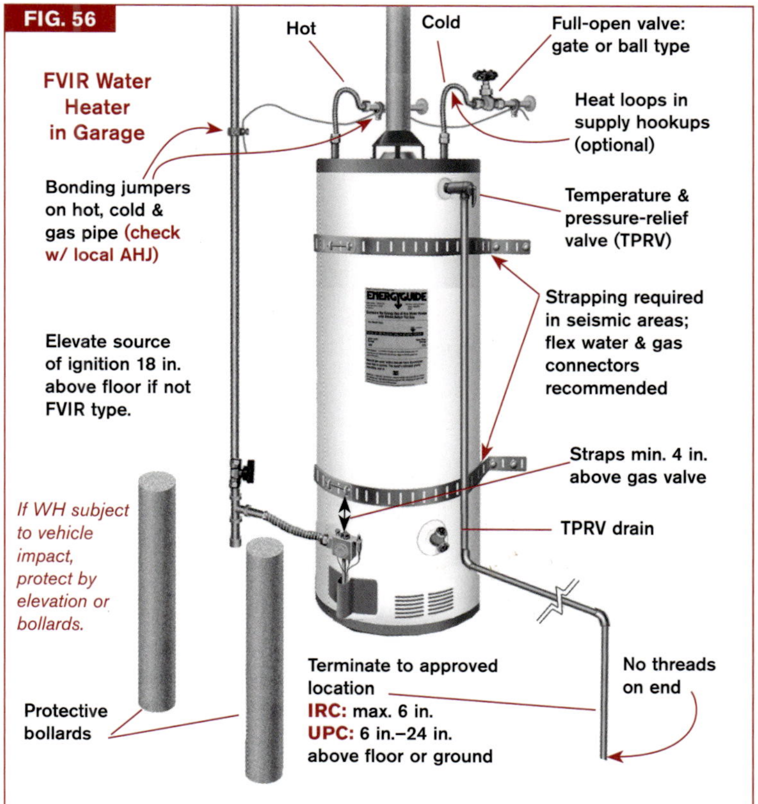

Flammable materials should not be stored near water heaters, even if the water heater is an FVIR type.

Tankless (On Demand) Water Heaters **F57**

	21 IRC	21 UPC
☐ Size AMI to meet demand	2005.1	501.1
☐ Size gas line to max. Btu input rating	2413.1	1215.2
☐ Vent AMI, Category III or IV typically required	2005.1	509.1.1
☐ PRV AMI, PRV size at appliance input rating or above	2804.2&3	504.6
☐ No venting in common unless AMI	2427.10.4	509.10.3.2
☐ Direct-vent water heater vent clearances per **F105,T51**	2427.8	T509.8.2

FIG. 57

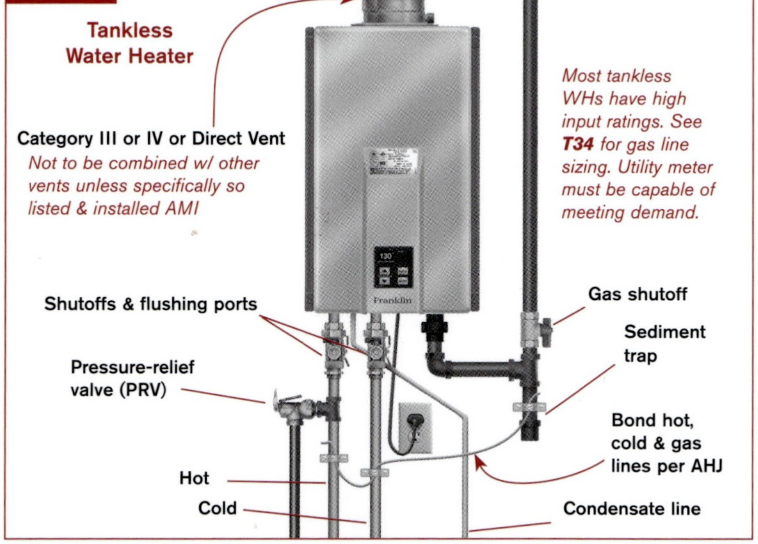

FIG. 58

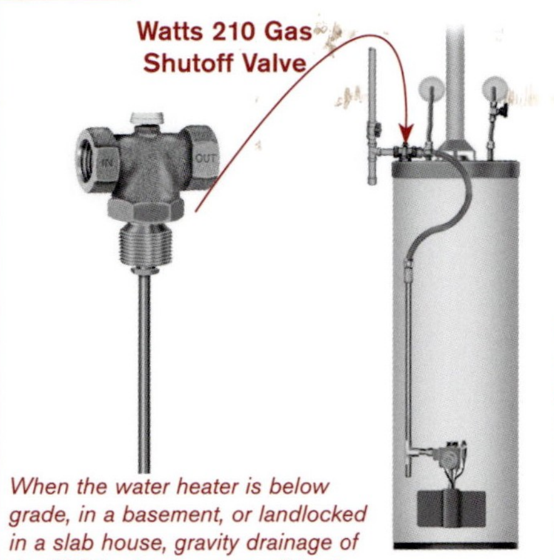

When the water heater is below grade, in a basement, or landlocked in a slab house, gravity drainage of the TPRV valve is not possible. A Watts 210 valve can be the solution.

FIG. 59

*The temperature-sensing bulb of the valve replaces the TPRV or goes in the upper portion of the tank. The gas piping runs through the valve & will shut off the gas if the temperature is excessive. A separate pressure-relief valve (**F59**) must be installed in the water piping & must drain by gravity to an approved location.*

Required Pans & Drain

	21 IRC	21 UPC
☐ Watertight corrosion-resistant pan required for WHs located where leakage could cause damage (UPC: to framing) **F61**	2801.6	507.5
☐ Pan min. 1½ in. deep **F61**	2801.6.1	507.5
☐ Pan not required under tankless water heater	2801.6	local
☐ Plastic pans under gas water heaters require flame spread index ≤25 & smoke-developed index ≤450	2801.6	n/a
☐ Min. ¾-in.-diameter drain required EXC **F61**	2801.6.1	507.5
• Not for replacement WH if none at existing pan	2801.6.1	n/a
☐ PVC not OK for pan drain piping	2801.6.1	n/a
☐ Pan drain required to end in indirect waste or outdoors 6 in. to 24 in. above grade (UPC: to any approved location)	2801.6.2	507.5

Temperature & Pressure-Relief Valves (TPRVs)

	21 IRC	21 UPC
☐ All WHs require pressure-relief device **F56,57,59–63**	2804.1	504.4
☐ All WHs require temperature-limiting device **F56,58,60,61**	2804.1	504.5
☐ IRC: devices may be combination TPRV (UPC: mandatory for storage-tank water heaters) **F56,60,61**	2804.1(2)	608.3
☐ Temp probe in top 6 in. of tank (UPC: AMI) **F56,58,61**	2804.4	504.5
☐ Device must open at 150 psi OR 210°F	2804.3&4	608.4
☐ Watts 210 also requires PRV **F58,59**	2804.1	504.6

FIG. 60 Combination Temperature & Pressure-Relief Valve

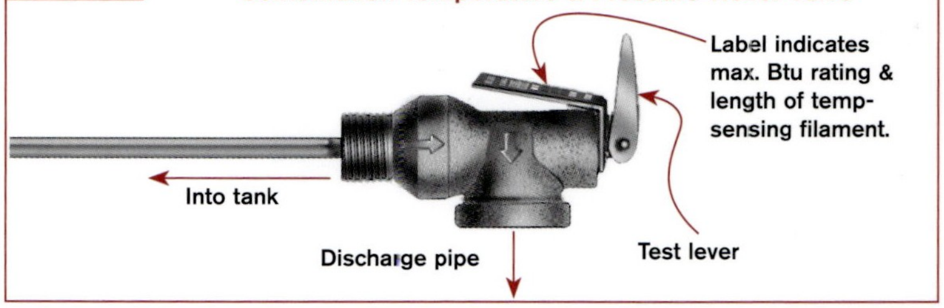

TPRV Drain Piping

	21 IRC	21 UPC
☐ No shutoff or check valves before or after TPRV	2804.6	504.6
☐ Must discharge through air gap in same room as WH	2804.6(2)	n/a
☐ Drain min. ¾ in. & ≥ valve outlet EXC	2804.6.1(3)	608.5(1)
• PEX or PE-RT min. 1 size larger if insert fittings used	2804.6.1(14)	n/a
• Secure end of PEX or PE-RT	2804.6.1(14)	n/a
☐ Material rated for temperature & in accord w/ ASME A112.4.1 (typically any material allowed for water distribution)	2804.6.1(13)	608.5(2)
☐ Drain must end outdoors or to indirect waste	2804.6.1(5)	608.5(3)
☐ May not drain to crawlspace	2803.6.1(6&7)	608.5(3)
☐ Terminate min. 2× pipe diameter max. 6 in. from ground or receptor (UPC: 6 in. min., 24 in. max.)	2804.6.1(10)	608.5(3)
☐ Must not discharge in a manner that could cause injury	2804.6.1(6)	608.5(4)
☐ Drain by gravity; cannot run uphill or be trapped	2804.6.1(8&9)	608.5(3&5)
☐ No threads on end of pipe	2804.6.1(11)	608.5(6)
☐ Each TPRV piped independently, may not be shared w/ condensate drain or relief valves of other systems	2804.6.1(4)	608.5(7)
☐ IRC: May discharge into pan (UPC: Prohibited) **F61**	2804.5.1	Ø 608.5(7)
☐ No kinks or restrictions in pipe	2804.6.1(3)	608.5(1)
☐ Terminate in readily observable location	2804.6.1(7)	608.5(8)
☐ End of pipe must be pointing down	n/a	608.5(1)
☐ Protect from freezing (terminate through air gap to indirect receptor located in a heated space)	2804.6.1(2)	608.5(5)

FIG. 61

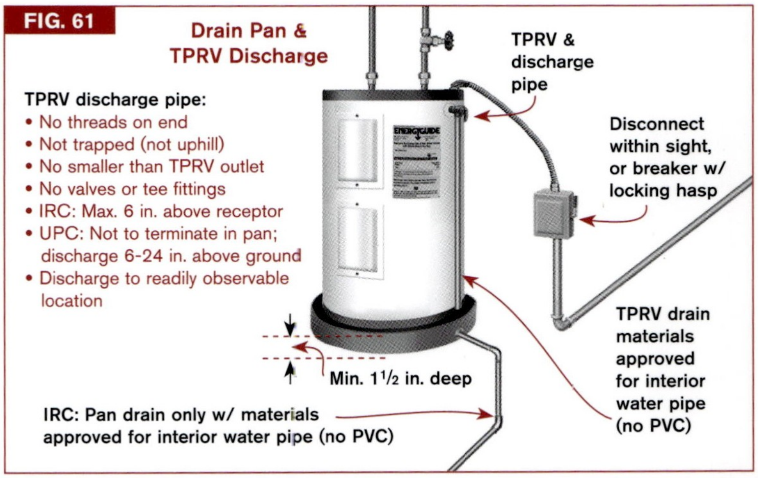

BOILERS & HYDRONICS

Modern high-efficiency boilers can be used for both hydronic heating systems and indirect-fired water heaters. Distribution can be through radiators, baseboard convectors, radiant in-floor tubing, or duct heaters. Valves, backflow preventers, drain piping, pipe support, and other items are also governed by plumbing codes.

Boilers

	21 IRC	21 UMC
☐ Oil-burning boilers require L&L to UL 726	2001.1.1	1002.2.1
☐ Installer to supply control diagram & operating manual	2001.1	1012.1
☐ Must be securely anchored to structure	1307.2	1001.5
☐ Clearance per L&L	2001.2	MFR
☐ Boilers require pressure & temperature gauges **F63**	2002.2	1003.3
☐ Steam boilers require sight-glass & pressure gauge	2002.3	1003.3
☐ Shutoff valves required in supply & return piping **F63**	2001.3	1212.3
☐ Low-water cutoff control required EXC	2002.5	1008.1
• Coil-type or forced-circulation boiler w/ flow sensor	2002.5X	1008.1
☐ PRV required **F63**	2002.4	1005.1
☐ PRV drain piped to within 18 in. of floor or receptor	2002.4	Ø
☐ Discharge piping—see rules for WH (*p. 30*)	n/a	1005.2

Dual-Purpose Water Heaters

	21 IRC	21 UMC
☐ Water heaters used for space heating & domestic HW L&L for the purpose & installed AMI **F62**	2004.1 & 2448.2	1203.2
☐ Tempering valve must limit potable water to 140°F **F62**	2803.2	1207.3.1

Expansion Tanks

	21 IRC	21 UMC
☐ Expansion tank required for every hydronic system **F63**	2003.1	1209.1
☐ Expansion tank must be rated for system pressure	2003.1.1	1209.1
☐ Tank test pressure 2½× allowable working pressure	2003.1.1	1004.3
☐ Min. tank capacity per **T40**	2003.2	1004.4

Valves

	21 IRC	21 UMC
☐ PRR required on makeup water feed **F63**	MFR	1214.4
☐ PRR required on low-pressure side	2101.23	1214.4
☐ Shutoff valve required both sides of PRRs	2101.22.4[10]	1212.5
☐ Air separation device required in hydronic systems	MFR	1214.7
☐ Isolation valves required at air-removal devices or vents	MFR	1212.11

10. This rule had formerly applied in the IRC only to ground-source heat pump systems.

Hydronic Piping: General

	21 IRC	21 UMC
☐ Support piping to avoid strain (*p. 6*)	2101.9	1210.3
☐ Install piping, valves, fittings & connections AMI (UMC: per applicable standards)	2101.24[11]	1210.1
☐ PEX & PE-RT require oxygen barrier in closed systems	n/a	1210.4
☐ Pressure test min. 1.5×design pressure min. 100 psi	2101.10	1205.2
☐ RP required to protect potable water if additives used	2101.3	1202.2
☐ Pipe bends min. 6× pipe diameter or AMI	MFR	1211.1
☐ If potable & nonpotable piping installed, provide identification on each piping system	2901.2	1204.1-5
☐ Indicate flow directions on system **F63**	n/a	1204.6

Embedded Piping (Radiant Heating)

	21 IRC	21 UMC
☐ Plastic pipe rated min. 80 psi at 180°F	2103.1	1221.2
☐ Cu tubing joints brazed not soldered	2103.3(2)	1221.2.2
☐ Tubing embedded in concrete min. 2 in. below surface	n/a	1217.5
☐ Min. R-5 insulation under poured concrete radiant system in soil contact & R-5 on vertical slab edges on grade	2103.2	1217.5.2

TABLE 40 MIN. EXPANSION TANK CAPACITY[A] IRC T2003.2[B] & UMC T1004.4(2)[C]

System Volume (gal)	Diaphragm Tanks	Non-Diaphragm Tanks	System Volume (gal)	Diaphragm Tanks	Non-Diaphragm Tanks
10	1.0	1.5	90	7.5	13.5
20	1.5	3.0	100	8.0[D]	15.0
30	2.5	4.5	200	17	30
40	3.0	6.0	300	25	45
50	4.0	7.5	400	33	60
60	5.0	9.0	500	42	75
70	6.0	10.5	1000	83	150
80	6.5	12.0	2000	165	300

A. Based on average water temperature of 195°F, fill pressure 12 psig, max. operating pressure 30 psig.
B. Volumes ≤100 gal. based on IRC T2003.2.
C. Volumes ≥100 gal. based on UMC T1004.4(2).
D. UMC table 1004.(2) gives this value at 9.0 gal. Both codes agree on 15.0 for non-diaphragm tanks.

11. Section 2101 (hydronic piping systems) was greatly expanded and now includes many specific requirements for joints and connections for different types of pipe and tubing.

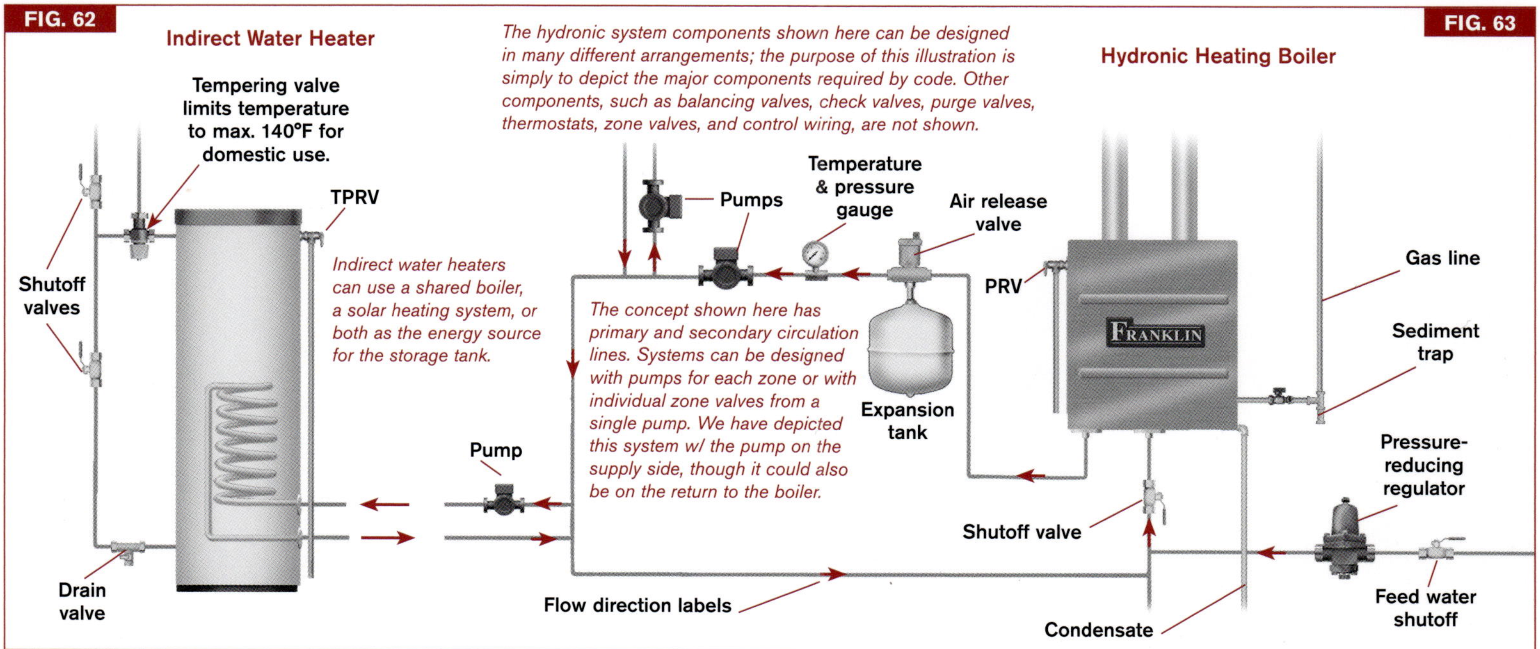

FIXTURES

Faucets, toilets, and showers must conform to federal water conservation regulations. Many states and local jurisdictions have stricter conservation standards.

General	21 IRC	21 UPC
☐ Max. water consumption of fixtures per **T41**	2903.2	See **T41**
☐ Smooth, impervious & no concealed fouling areas **F64**	2701.1	401.2
☐ Watertight seal required between fixtures & walls or floors (e.g., caulk base of toilet)	2705.1(3)	402.2
☐ Rigidly secure fixtures to walls/floors	2705.1(1&2)	402.3&4
☐ Valves required on each fixture supply EXC	2903.9.3	606.5
• Tubs & showers (UPC: fixtures w/o slip joints)	2903.9.3	606.5
• Shutoff may be at manifold **F40**	2903.8.4	606.5
☐ Hot on left, cold on right when facing outlet EXC	2722.2	417.5
• Single-handle controls—orientation per control labels	2722.2X	417.5
☐ Sinks & lavs require drain strainer (UPC: lav stopper OK)	2702.1	407.5&420.4
☐ Tailpiece min. 1¼ in. for lavs & bidets	2703.1	T702.1
☐ Tailpiece min. 1½ in. for other residential fixtures	2703.1	420.4
☐ Floor drains require removable strainers	2719.1	418.2

Testing Fixtures		
☐ After fixtures set, fill traps & prove gastight or watertight	2503.2	712.1
☐ Water test: Fill each fixture, drain & visually inspect	2503.2(1)	712.1
☐ Gas test w/ peppermint oil: Introduce 2 oz. oil, 10 quarts hot water, seal vent terminals, check for odor at traps	2503.2(2.2)	local

FIG. 64

Concealed Fouling Areas

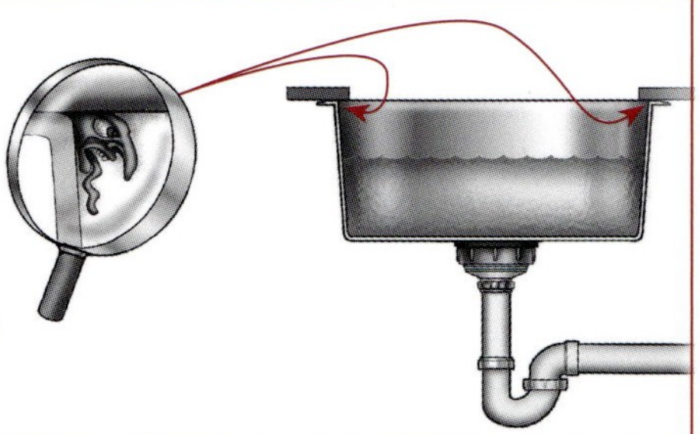

No part of a fixture should be concealed such that it cannot be cleaned. Sinks secured to the underside of a drainboard often harbor a neglected fouling surface.

TABLE 41	MAX. FLOW RATES FOR PLUMBING FIXTURES IRC T2903.2 & UPC 407.2, 408.2, 411.2, 420.2
Plumbing Fixture or Fixture Fitting	**Max. Flow Rate**
Lavatory faucet	2.2 gpm at 60 psi[A]
Shower head[B]	2.5 gpm at 80 psi
Sink faucet	2.2 gpm at 60 psi
Water closet	1.6 gal. per flush

A. Private only; max. flow rate for public lav faucet is 0.5 gpm in the IPC & UPC.
B. Handheld shower sprays are also considered shower heads.

FIG. 65

Slip-Joint Access

Slip joints must be accessible for service. Access opening should be at least 12 in. by 12 in. to allow proper service access. The opening can be in a ceiling or a wall.

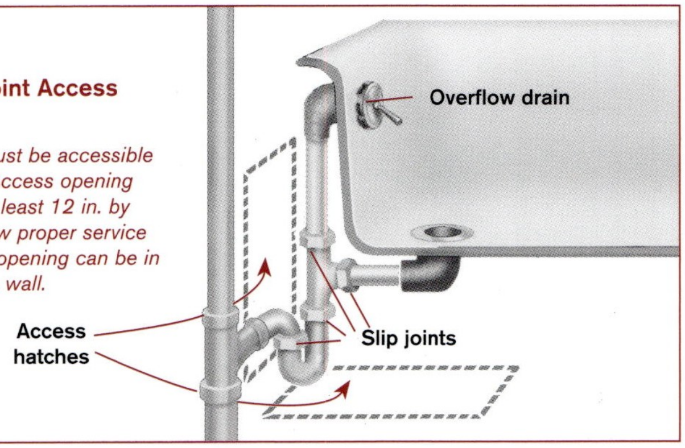

The supply to a dishwasher is required to have backflow protection in the form of an air gap complying with ASME A112.1.3 or A112.1.2. The supply air gap protects the potable water contamination by the dishwasher. The drainage air gap protects the clean dishes from being contaminated by a drain line backup.

Kitchen Sinks & Fixtures	21 IRC	21 UPC
☐ Sink min. outlet 1½ in. diameter	2714.1	T702.1
☐ OK for sink, DW & disposer on same 1½-in. trap	2717.2	T702.1
☐ 2-in. drain required for sink downstream of trap	n/a	T702.1
☐ DW supply requires air gap or integral backflow device	2717.1	602.3
☐ DW may discharge (UPC: Through air-gap fitting) directly to branch wye tailpiece on KS or food waste disposer head **F66,67**	2717.2	414.3
☐ Secure DW drain hose to underside of counter **F66**	2717.2	n/a
☐ Air-gap fitting required for DW drain **F67**	n/a	807.3
☐ Air-gap fitting critical level above FLR of sink	n/a	807.3
☐ No DW connection to discharge side of disposer	MFR	414.3
☐ Reverse osmosis systems to recognized standards	2909.2	611.1
☐ Reverse osmosis system discharge requires air gap **F68**	2909.2	611.2
☐ No saddle fittings or tapping/drilling of drain line	3003.2	310.2

FIG. 66

Dishwasher Drain with High Loop

FIG. 67

Drainage Air-Gap Device

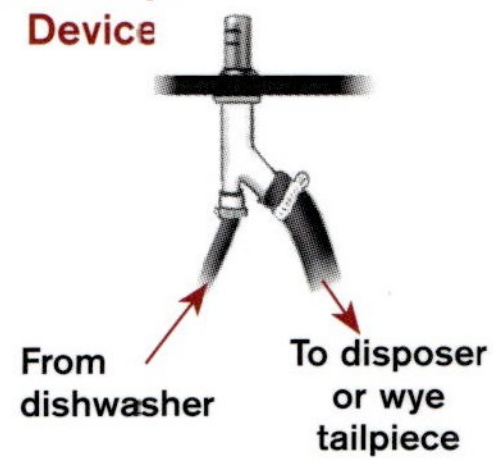

The UPC & many dishwasher manufacturers require that the dishwasher drain connect to the plumbing drainage system through an air-gap device.

FIG. 68

Reverse Osmosis Water Treatment

Reject water line must discharge to drainage system through an air gap.

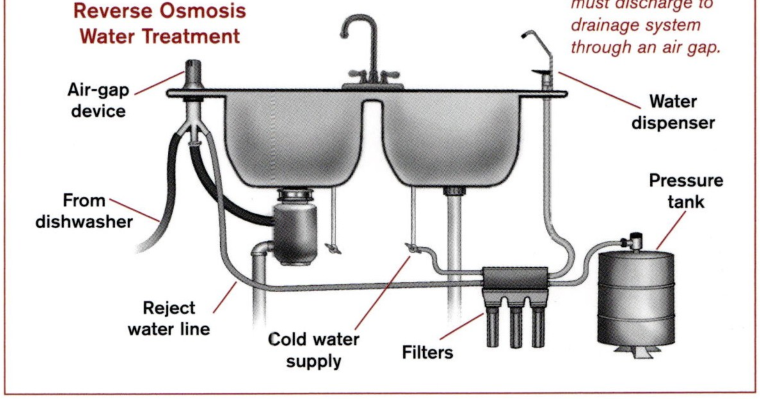

Tubs & Whirlpools *(also see Code Check Electrical)*	21 IRC	21 UPC
☐ Slip joints accessible, min. 12-in. × 12-in. access **F65**	2704.1	402.10
☐ Tub waste opening to crawlspace requires rodent screen	n/a	312.12.3
☐ Over-rim spigot min. vertical air gap 2× outlet diameter	T2902.3.1	T603.3.1
☐ Tubs, whirlpools require temp-limiting device at max. 120°F, water heater thermostat is not acceptable method	2713.3	409.4
☐ Whirlpool pump accessible & above trap crown weir	2720.2	409.6
☐ Pump access ≥12 × 12 in., 18 × 18 in. if >2 ft. from access	2720.1	n/a

FIG. 69

Bathroom Fixture Layout

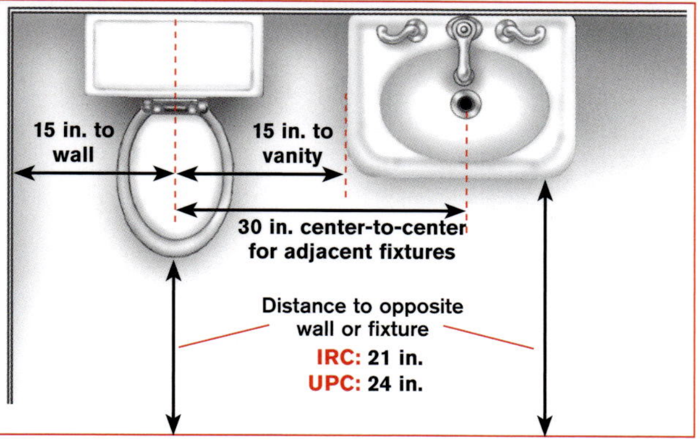

Showers

	21 IRC	21 UPC
☐ Min. area 900 sq. in. (UPC: 1,024 sq. in.) **F70** EXC	2708.1	408.6
• Fold-down seats protruding into space must allow the min. 900-sq.-in. area w/ seat in folded-up position	2708.1X1	n/a
☐ Must be able to encompass 30-in.-diameter circle from top of threshold to point 70 in. above drain outlet EXC	2708.1	408.6
• Shower heads, valves, grab bars & soap dishes allowed to protrude into required min. space	2708.1	408.6
• 25-in. cross section OK if area ≥1300 sq. in.	2708.1X2	n/a
• Not required if min. 30-in. × 60-in. enclosure	n/a	408.6X2
☐ Shower walls nonabsorbent to min. 72 in. above drain	307.2	local
☐ If threshold provided, height min. 1 in. below top of shower receptor membrane & min. 2 in. max. 9 in. above top of drain **F71**	2709.1	408.5
☐ If no threshold, adjacent floor considered a wet location	local	408.5
☐ Door min. 22-in.-wide clear egress opening **F70**	2708.1.1	408.5
☐ Door must open outward **F70**	2708.1	local
☐ Finished floor slope min. ¼ in./ft. (UPC ⅛ in.) max. ½ in./ft.	2709.1	408.5
☐ Secure shower valve, head/riser to permanent structure	2708.3	408.10
☐ Shower head cannot discharge directly at entrance	local	408.9
☐ Listed anti-scald/pressure balance valve 120° max.	2708.4	408.3
☐ Min. 2-in. drain outlet (IRC 1½-in.), strainer of equivalent area	2708.2	408.4

FIG. 70

Shower Receptor

Overall net area ***IRC****: 900 sq. in.* ***UPC:*** *1024 sq. in*

Soap dishes, grab bars, shower heads & valves allowed in 30-in.-wide space.

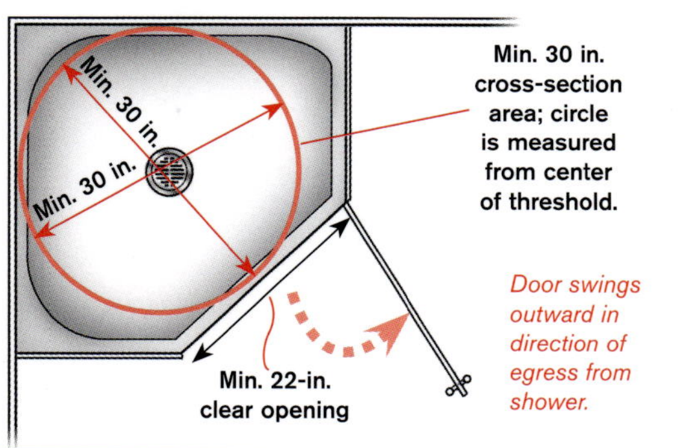

Shower Pan & Liner

	21 IRC	21 UPC
☐ Site-built liner materials conform to approved standards	2709.2	408.7
☐ Slope underlayment ¼ in./ft. to weep holes **F71**	2709.3	408.7
• Liner min. 2 in. above dam or threshold (UPC: 3 in.) **F71**	2709.2	408.7
☐ Pan liner plastic AMI or 3 layers hot mop type 15 felt	2709.2	408.7
☐ Special attention to hot mop corner installation; extend 4 in. all directions from corner	2709.2.3	408.7
☐ PVC & CPE sheet lining cemented AMI	2709.2.1&2	408.7.1&2
☐ Weep holes at drain required & must remain clear **F71**	2709.4	408.7
☐ No fasteners in liner <1 in. above finished threshold	2709. 3	408.7
☐ Roll over top of rough threshold (no penetrations through top) & fasten to outside edge **F71**	2709.3	408.7
☐ Watertight connection between liner and drain flange	2709.4	408.7

Shower Testing

	21 IRC	21 UPC
☐ Pan leak test min. 2 in. water measured at threshold, min. 15 minutes (UPC: Water level ≥ rough threshold height)	2503.6	408.7.5
☐ If no threshold (curbless), build temporary threshold	2503.6	local
☐ Plug pipe w/ test balloon below weep holes	2503.6	408.7.5
☐ Must hold water min. 15 minutes w/ no leakage	2503.6	local

FIG. 71 Site-Built Shower Pan

Lining material min. 2 in. (UPC: 3 in.) above threshold or dam. Recess from face so as not to interfere with wall covering.

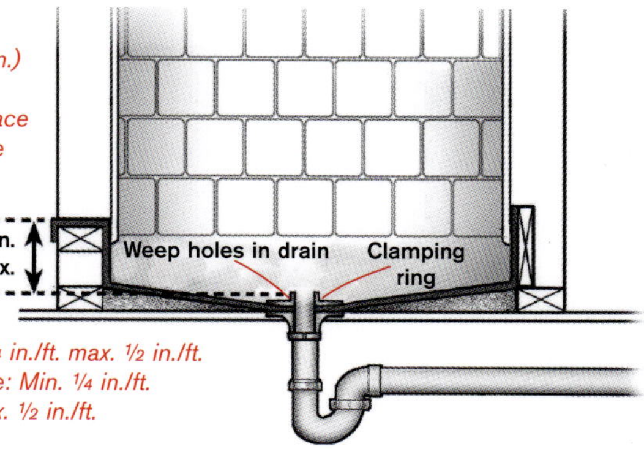

Liner slope: Min. ¼ in./ft. max. ½ in./ft.
Finished floor slope: Min. ¼ in./ft. (⅛ in./ft. UPC) max. ½ in./ft.

Toilets & Bidets

	21 IRC	21 UPC
☐ Secure floor flange to drainage connection & to floor w/ corrosion-resistant fasteners	2705.1(1)	402.6.2
☐ Install flanges AMI w/ no concealed fouling area **F72**	2701.1	402.6.1
☐ No offset closet flanges for floor-mounted back outlet WCs	n/a	402.6.3
☐ 15-in. min. clearance from center toilet (WC) or bidet to side wall, vanity, or outer rim of adjacent fixtures **F69**	2705.1(5)	402.5
☐ 21 in. min. (UPC : 24 in.) clear in front of lav or toilet **F69**	2705.1(5)	402.5
☐ Toilet or bidet min. 30-in. separation center to center	2705.1(5)	402.5
☐ Ballcock critical level ≥1 in. above overflow pipe **F73**	2902.4.1	603.5.2
☐ Joint between fixture & floor or wall to be watertight	2705.1(3)	402.2
☐ Bidet fixture supply requires air-gap vacuum breaker	2721.1	410.2
☐ Max. discharge temp at bidet 110°F	2721.2	410.3

FIG. 72 Toilet Flanges (Closet Rings)

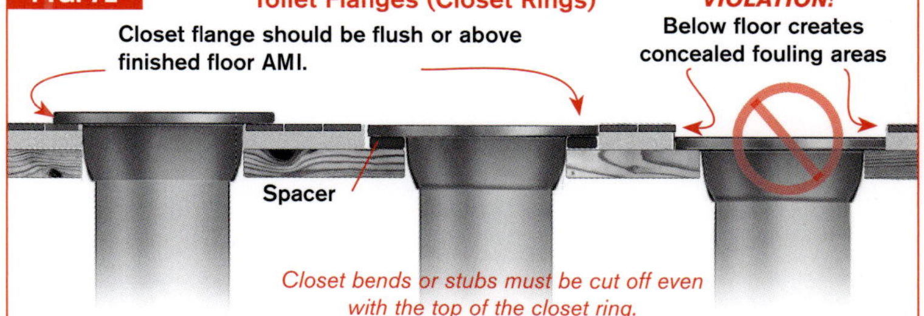

FIG. 73 Flush Tank Fill Valves

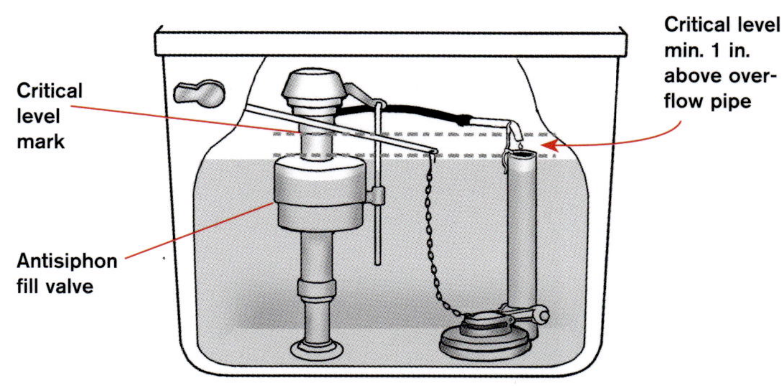

Clothes Washers & Laundry Sinks (Trays)

	21 IRC	21 UPC
☐ Standpipe ≥18 in. & ≤42 in. (30 in. UPC) above trap **F74**	2706.1.2	804.1
☐ Must drain through air break (no pressurized waste)	2718.1	805.1
☐ No trap below floor	n/a	804.1
☐ Trap ≥6 in. & ≤18 in. above floor **F74**	n/a	804.1
☐ LT may drain into washer standpipe within 30 in. if standpipe min. 30 in. above trap weir & above FLR of LT **F75**	2706.1.2.1	local

FIG. 74

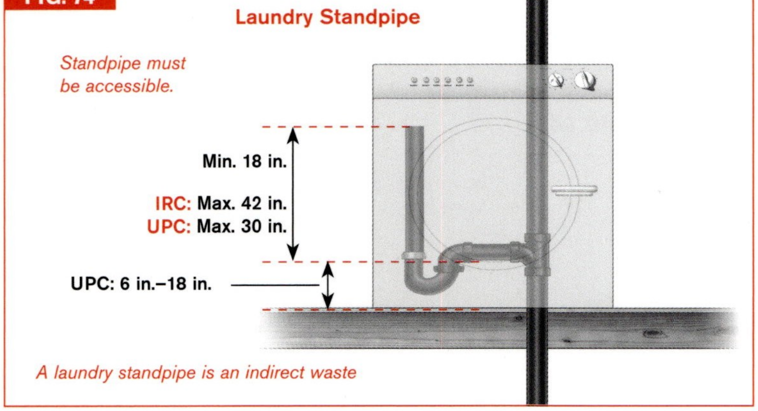

FIG. 75

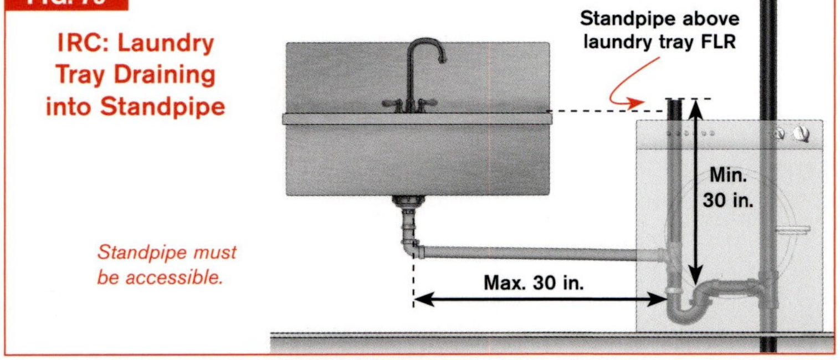

CLOTHES DRYERS

Houses built prior to adoption of the 1996 NEC often have 3-hole 240-volt dryer receptacles, where the dryer neutral is also its equipment ground. Dryers installed in such receptacles should have their internal bonding jumper connected. In newer installations with 4-wire cords the bonding jumper should be removed. See *Code Check Electrical* for further information.

IRC Clothes Dryer Exhaust—General

	Electric	Gas
☐ Exhaust AMI w/ dryer MFR	1502.1	2439.1
☐ Exhaust must be independent of other systems	1502.2	2439.1
☐ Exhaust must convey moisture to outdoors EXC	1502.2	2439.1
• L&L condensing (ductless) dryers	1502.2X	2439.7.6X
☐ Must convey combustion products to outdoors	n/a	2439.1
☐ Exhaust must terminate outside building min. 3 ft. from any opening into building (5 ft. recommended to AC coils)	1502.3	1502.3
☐ Termination requires backdraft damper, no screens **F76**	1502.3	2439.3
☐ Closet locations require min. 100-sq.-in. makeup air opening	n/a	2439.5.1

TABLE 42 DRYER FITTING EQUIVALENT LENGTH IRC T1502.4.6.1 & T2439.7.4.1

Fitting Radius	Equivalent Length	
	45° Elbow	90° Elbow
4 in. mitered	2 ft. 6 in.	5 ft.
6 in. smooth	1 ft.	1 ft. 9 in.
8 in. smooth	1 ft.	1 ft. 7 in.
10 in. smooth	9 in.	1 ft. 6 in.

IRC Clothes Dryer Exhaust Ducts

	Electric	Gas
☐ If dryer space provided, exhaust duct required EXC	1502.4.8	2439.7.6
• Listed condensing dryer installed prior to occupancy	1502.4.8X	2439.7.6X
☐ Duct must be smooth metal 4 in. nominal diameter	1502.4.1	2439.7.1
☐ Duct requires support at max. 12 ft. (4 ft. gas) intervals	1502.4.2	2439.7.2
☐ Male end at joints must point in direction of flow	1502.4.2	2439.7.2
☐ Ducts concealed in cavities shall not be deformed	1502.4.2	2439.7.2
☐ Screws & fasteners not to project > ⅛ in. into duct	1502.4.2	24397.2
☐ Transition ducts (connectors) L&L to UL 2158A	1502.4.3	2439.7.3
☐ Transition ducts (connectors) max. 8 ft., not concealed	1502.4.3	2439.7.3
☐ Shield plates if duct <1¼ in. from framing surface	1502.5	2439.6

IRC Clothes Dryer Exhaust Duct Length

	Electric	Gas
☐ Option 1: Max. 35 ft. minus bends per **T42**,**F76**	1502.4.6.1	2439.7.4.1
☐ Option 2: AMI—provide installation instructions to BO	1502.4.6.2	2439.7.4.2
☐ Option 3: Lengths >35 ft. AMI of power ventilator	1502.4.6.3	2439.7.4.3
☐ Lengths >35 ft. require sign within 6 ft. of connection stating equivalent length of exhaust duct	1502.4.7	2439.7.5
☐ Power ventilators L&L to UL 705, installation AMI	1502.4.4	2439.4
☐ Booster fans (not listed to UL 705) prohibited	1502.4.5	2439.4

FIG. 76

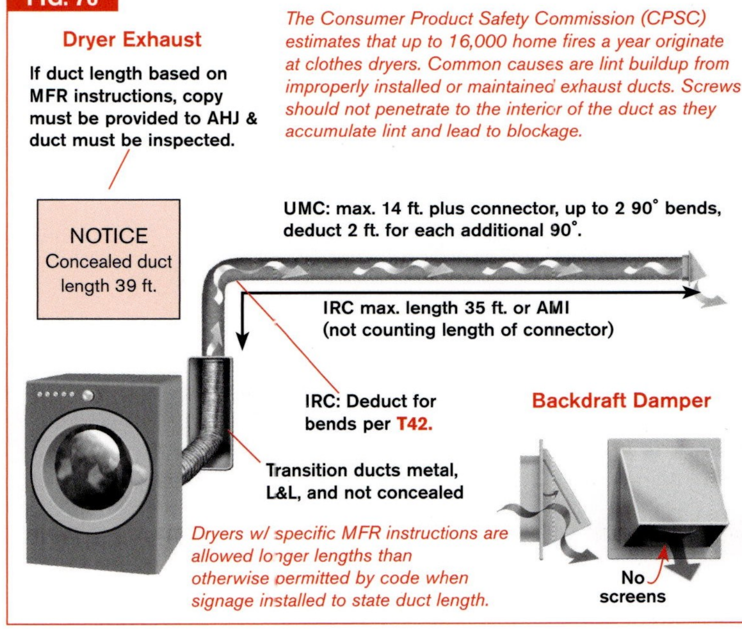

UMC Clothes Dryer Exhaust

	21 UMC
☐ Exhaust must be independent of other systems	504.4
☐ Duct must be rigid metal 4 in. nominal diameter	504.4 & 504.4.2
☐ Male end at joints must point in direction of flow	504.4
☐ Screws & fasteners not to project into duct	504.4
☐ Terminate outside w/ backdraft damper, no screens **F76**	504.4
☐ Exhaust must terminate outside building min. 3 ft. from any opening into building (5 ft. recommended to AC coils)	504.4 & 502.2.1
☐ Transition ducts (connectors) L&L to UL 2158A **F76**	504.4
☐ Transition ducts (connectors) max. 6 ft., AMI & not concealed **F76**	504.4.2.2
☐ Closet locations require min. 100-sq.-in. makeup air opening	504.4.1
☐ Duct length total combined vertical & horizontal length 14 ft. EXC	504.4.2.1
• AMI & approved by AHJ	504.4.2.1
• AMI per duct power ventilator	504.4.2.1X
☐ Deduct 2 ft. from allowed length for each elbow in excess of 2	504.4.2.1
☐ Power ventilators require L&L to UL 705, installation AMI	504.4.2.3

VENTILATION & EXHAUST SYSTEMS

In general, building codes tell us ***when*** to provide ventilation for interior spaces, while mechanical and energy codes tell us ***how*** to provide it. Proper ventilation is necessary to maintain a healthy indoor environment. *ASHRAE 62.2, Ventilation and Acceptable Indoor Air Quality in Residential Buildings*, is the standard referenced by many energy codes, and the same material appears in abbreviated form in IRC section 1103.6 and 1505. Check with your local jurisdiction to determine the standards in your area. One purpose of whole-house ventilation is to dilute contaminants from materials such as volatile organic compounds (VOCs) found in furnishings and building products. Localized exhaust removes contaminants from specific sources, such as kitchens and baths.

Whole Building Ventilation — ASHRAE 62.2

- ☐ Mechanical exhaust, supply, or combination system required ____ 4.1
- ☐ Min. ventilation rate must comply w/ **T43** EXC ____ 4.1.1
  - • In Climate Zone 1 or 2, local AHJ determines that window operation provides sufficient ventilation, no mechanical cooling is present & thermally conditioned for human occupancy < 876 hr. per year ____ 4.1.1X
- ☐ Measured infiltration rate can be used as credit to required ventilation ____ 4.1.2
- ☐ **T43** assumes 2 persons in studio or 1-bedroom unit & 1 person per each additional bedroom; add 7.5 cfm per person for higher densities ____ 4.1.3
- ☐ 20% reduction to **T43** allowed w/ improved air filtration ____ 4.1.4
- ☐ Local exhaust fans count to required whole-house continuous ventilation ____ 4.2
- ☐ Outdoor air duct to return of air handler can supply ventilation if return air temp per MFR ____ 4.2
- ☐ Quantity of air can be based on equipment MFR install instructions ____ 4.3
- ☐ ON-OFF control w/ ready access & text indicating its purpose EXC ____ 4.4
  - • Not required to be readily accessible in multifamily dwelling units ____ 4.4X

TABLE 43 — MIN. VENTILATION RATES IN CFM ◆ ASHRAE 62.2 T4.1A

Floor Area (sq. ft.)	Number of Bedrooms				
	1	2	3	4	5
< 500	30	38	45	53	60
501–1000	45	53	60	68	75
1001–1500	60	68	75	83	90
1501–2000	75	83	90	98	105
2001–2500	90	98	105	113	120
2501–3000	105	113	120	128	135
3001–3500	120	128	135	143	150
3501–4000	135	143	150	158	165
4001–4500	150	158	165	173	180
4501–5000	165	173	180	188	195

Kitchen Exhaust — ASHRAE 62.2 — 21 UMC

- ☐ Mechanical exhaust required for each kitchen ____ 5.1 — 405.4[12]
- ☐ Exhaust may be intermittment (on demand) or continuous ____ 5.1 — 405.3&4[12]
- ☐ Kitchen exhaust rate min. 100 cfm for range hoods ____ T5.1 — 405.4.1[12]
- ☐ Other kitchen exhaust fans min. 300 cfm ____ T5.1 — 405.4.1[12]
- ☐ Downdraft appliance exhaust fans min. 300 cfm ____ T5.1 — 405.4.1[12]
- ☐ Continuous kitchen exhaust rate min. 5 ACH ____ T5.2 — 405.4.1[12]
- ☐ Duct min. diameter per **T44** ____ T5.3[13] — n/a

TABLE 44 — PRESCRIPTIVE DUCT SIZES (IN.) ◆ ASHRAE 62.2 T5-3

Fan cfm	50	80	100	125	150	175	200	250	350	400	450	700	800
Rigid duct	4	5	5	6	6	7	7	8	9	10	10	12	12
Flex duct[A]	4	5	6	6	7	7	8	8	9	10	NP	NP	NP

A. Flex duct must be fully extended. Elbows to have min. bend radius of one duct diameter.

12. UMC section 405 is new in the 2021 edition. In addition to the requirements shown here, it requires whole house ventilation & natural ventilation similar to the previous page & to the building code.
13. The 2016 edition of this ASHRAE table also included maximum lengths.

Other ASHRAE Requirements — ASHRAE 62.2

- ☐ Label controls as to function unless obvious by location ____ 6.2
- ☐ Door from attached garage to house weatherstripped ____ 6.5.1
- ☐ Space conditioning w/ equipment outside pressure boundary max. duct leakage 6% using Method D of ASTM E1554 - also see ***p.36*** ____ 6.5.2
- ☐ Habitable spaces require ventilation ≥4% (min. 5 sq. ft.) of floor area ____ 6.6.1
- ☐ Utility/WC rooms ventilation ≥4% (min. 1½ sq. ft.) of floor area EXC ____ 6.6.2
  - • Utility rooms w/ dryer exhaust duct ____ 6.6.2X
- ☐ Central furnace or AC system filter min. efficiency MERV 6 ____ 6.7
- ☐ Air inlets min. 10 ft. from contaminants such as plumbing vents ____ 6.8
- ☐ Carbon monoxide alarms required in each dwelling unit ____ 6.9

Sound Levels

- ☐ Whole building or continuous ventilation fans max. 1.0 sone ____ 7.2.1
- ☐ Bath exhaust fans max. 3 sones ____ 7.2.2
- ☐ Kitchen exhaust max. 3 sones at 1 or more settings ≥100 cfm EXC ____ 7.2.2
  - • Kitchen exhaust >400 cfm ____ 7.2.2

Bathroom Exhaust

- ☐ Mechanical exhaust required each bath w/ tub or shower ____ 5.1
- ☐ Min. exhaust rate 50 cfm intermittent or 20 cfm continuous ____ T5.1 & T5.2
- ☐ Controls manual or automatic w/ ON override ____ 5.2.1
- ☐ Duct min. diameter per **T44** ____ T5.3[13]

IRC Bathroom Exhaust & Ventilation — 21 IRC

- ☐ Bathrooms & toilet rooms require min. 3-sq.-ft. glazed openings w/ min. 50% openable for ventilation EXC ____ 303.3
  - • Not required if provided w/ fan direct to exterior **F77** ____ 303.3X
- ☐ Min. exhaust rate 50 cfm intermittent or 20 cfm continuous ____ T1505.4.4
- ☐ Fan must exhaust directly to outdoors ____ 1505.2
- ☐ Air may not be exhausted into attic, soffit, ridge, or crawlspace **F77** ____ 1501.1

FIG. 77

Bathroom Exhaust

Bath or laundry exhaust ducts are commonly used to comply with the requirement for continuous whole-building ventilation.

VIOLATION!

Ducts may not discharge into attics, soffits, or crawlspaces.

Exhaust air is vented to the outside to prevent mold or mildew from condensation.

Exhaust Ducts — ASHRAE 62.2

- ☐ Exhaust fans in separate dwelling units not to share a common duct ____ 7.3.1
- ☐ If >1 fan in a single dwelling unit shares a common duct, each fan requires a backdraft damper ____ 7.3.1
- ☐ Commonly ducted exhaust inlets in multiple units require continuously running exhaust fan downstream of inlets ____ 7.3.2

Exhaust Ducts & Openings — 21 IRC — 21 UMC

- ☐ Flex ducts min. 4 in. diameter ____ T1504.2 — 502.2.1
- ☐ Exhaust air not to be directed onto walkways ____ 303.5.2 — 502.2.1
- ☐ Exhaust air to terminate min 3 ft. from LL, 3 ft. from gravity air intake openings, operable windows, & doors, & 10 ft. from mechanical air intake openings ____ 303.5.2 — 502.2.1

(Note: ASHRAE 62.2 section 6.8X2 allows a bathroom/kitchen fan to terminate outdoors less than 3 ft. from an openable window to the bathroom/kitchen.)

- ☐ Exhaust & intake openings require screens ____ 303.6 — 502.1
- ☐ Screen mesh openings min. ¼ in. max. ½ in. ____ 303.6 — 502.1
- ☐ Exhaust ducts require backdraft dampers ____ 1103.6 — 504.1.1

Energy-Recovery Ventilators F78	21 IMC	21 UMC
☐ Non-ducted to comply w/ UL 1815, ducted to UL 1812	514.1	504.5
☐ Install per L&L & AMI	304.1	504.5

FIG. 78 Energy-Recovery Ventilator

Required in climate zones 7 & 8

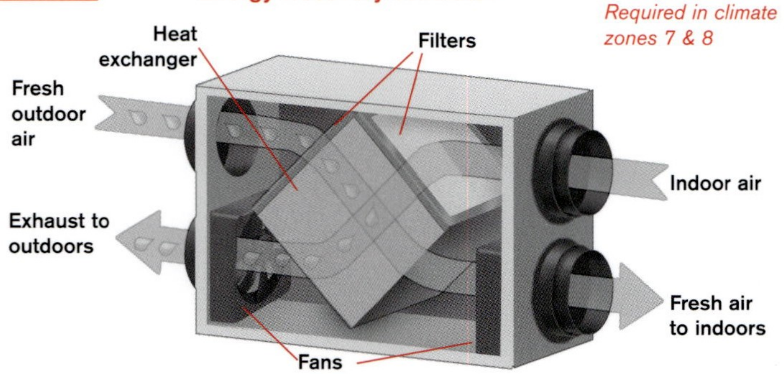

RANGES & RANGE HOODS

Freestanding Ranges	21 IRC	21 UMC
☐ L&L as household type—not commercial	1901.2 & 2447.2&3[14]	902.1,MFR
☐ Anti-tip bracket required AMI	MFR	MFR
☐ Vertical clearance to combustibles min. 30 in. F79 EXC	1901.1	920.3.2
• Clearances per L&L of range hoods or ovens F79	1901.1	920.3.2(3)
• 24 in. OK w/ metal hood or metal over ¼-in. millboard	n/a	920.3.2(1)
☐ Side clearance to combustibles AMI EXC	1901.2	920.3.1(1)
• 6 in. min. sides & rear for unlisted appliances	Ø	920.3.1(4)
Built-in Ranges		
☐ Install AMI	1901.1 & 2447.1	920.4.1
☐ Vertical clearance to combustibles min. 30 in. EXC F79	1901.1	920.4.2
• Lesser clearances per L&L and AMI F79	1901.1	920.4.2(3)
• 24 in. OK w/ metal hood or metal over ¼-in. millboard	n/a	920.4.2(1)
☐ Must be level	MFR	920.4.3
☐ Must be ducted to outdoors & have backdraft damper	1503.3	504.1.1
Range Hoods & Exhaust Ducts		
☐ Terminate min. 3 ft. from building openings	1504.3	502.2.1
☐ Gas ranges do not require vent to outdoors	2425.8(1)	802.2.1(1)
☐ Ducts smooth interior surface (not corrugated)	1503.3	504.3
☐ Ducts must be Zi steel, SS, or Cu EXC	1503.4	504.3
• Downdraft duct, PVC OK for portion under slab	1503.4X	504.3X
☐ Terminate outdoors—not in attic or crawlspace EXC	1503.3	504.1.1
• L&L ductless range hoods installed AMI	1503.3X	504.1X(1)
Hoods for Open-Top Broilers		
☐ Hood required & must extend as wide as broiler unit EXC	1503.2.1	922.3
• Hood not required for L&L units w/ integral exhaust	1503.2.1X	922.3
☐ Min. ¼-in. clearance between hood & underside of combustible or metal cabinets	1503.2.1	922.3
☐ Min. 24 in. from cooking surface to combustible materials	1503.2.1	922.3

FIG. 79 Range Clearances

Lesser clearances allowed for listed appliances per terms of listing

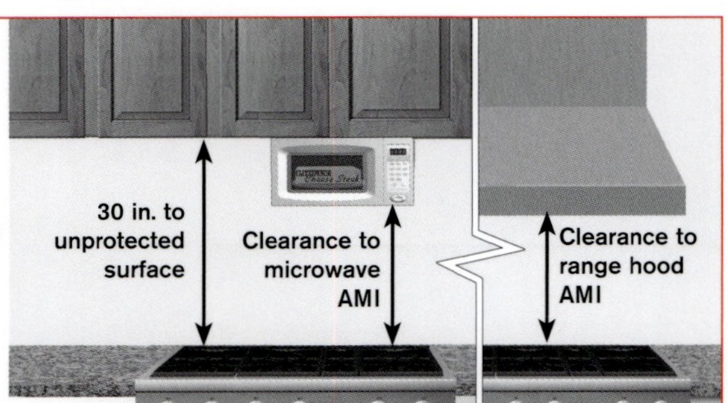

14. Commercial ranges were allowed under direction of licensed professional engineer. This rule has been removed, as commercial ranges have very different clearance requirements. Many manufacturers supply "commercial style" ranges listed as household type.

DUCTS

ACCA Manual D is the design standard for sizing residential duct systems and helps to ensure a balanced airflow throughout a building. Factory-made ducts and closure systems (tape) must comply with UL standards. SMACNA provides fabrication and support standards for sheet-metal ducting and installation standards for factory-made duct systems. The Air Duct Council (ADC; formerly the Air Diffusion Council) also provides standards for installation & support of factory-made flexible ducts. Standards for insulation and duct leakage testing may be superseded by local or state energy codes.

General	21 IRC	21 UMC
☐ Duct sizes per ACCA Manual D or other approved methods	1601.1	601.2X
☐ Factory-made ducts L&L per UL 181	1601.1.1(2)	603.4
☐ Max. 2 stories for vertical riser on factory-made duct	n/a	603.1.4
☐ Fireblock openings around ducts between floors	302.11(4)	local
☐ Vibration isolators max. 10 in. length	1601.2	602.5
☐ Garage ducts min. 26-ga sheet steel; no flex	302.5.2 & 1601.4.9	local
☐ No duct openings into garage	302.5.2	local
Underground Ducts		
☐ Metal ducts protected from corrosion in approved manner or completely encased in ≥2 in. of concrete	1601.1.2	603.11
☐ Ducts sealed, secured & tested prior to encasement	1601.1.2	603.11
☐ Ducts must be sloped to drain to drainage point (UMC: Min. ⅛ in./ft. back to main riser)	1601.1.2	603.11
☐ Factory-made ducts not allowed underground	1601.4.7	603.11

SMACNA Rigid Duct Support	ANSI/SMACNA 006
☐ Rigid horizontal ducts require support within 2 ft. of each elbow	5.1.1
☐ Support rigid horizontal ducts within 4 ft. of each branch intersection	5.1.1
☐ Rigid round duct max. support spacing 12 ft.	T5-2
☐ Round duct up to 24-in. diameter: Support straps 1 in. × 22-ga	T5-2
☐ Round duct up to 10-in. diameter: Min. support 12-ga wires	T5-2
☐ Min. two 12-ga or one 8-ga wire for round ducts 11 to 18 in. diameter	T5-2
☐ Min. two 10-ga wire for round ducts 19 in. to 24 in. diameter	T5-2

UMC Rigid Duct Support	21 UMC
☐ Support at each change of direction	603.7.1
☐ Riser ducts require support by metal straps or angles to the structure	603.7.1
☐ Rectangular ducts require support on 2 opposite sides of each duct attached by rivets, bolts, or screws at intervals in SMACNA T5-1	603.7.1.1
☐ Horizontal round ducts ≤40 in. diameter 1 hanger per 12-ft. interval	603.7.1.2
☐ Horizontal round ducts require tight-fitting circular bands around entire perimeter of duct at each support interval EXC	603.7.1.3
• Round ducts ≤10 in. diameter 18-ga Zi steel wire OK as support	603.7.1.4X
☐ Circular bands around ducts min. 1 in. wide	603.7.1.4

Joints & Seams	21 IRC	21 UMC
☐ Joints, seams & connections per SMACNA standards	1601.4.1	603.9
☐ Mechanically fasten ducts to plenums & flanges	1601.4.1	603.9
☐ Crimp joints contact lap min. 1 in. (UMC: 1½ in.)	1601.4.1	603.9
☐ Crimp joints fasten w/ min. 3 sheet-metal screws	1601.4.1	603.9
☐ Male end into adjoining duct in direction of flow	1601.4.2	603.9
☐ Flex duct fasteners must be per UL 181 B	1601.4.1	603.9.1
☐ Tape for flex ducts requires UL 181 B-FX marking F81	1601.4.1	603.9.1
☐ Mastic for flex ducts requires UL 181 B-M marking	1601.4.1	603.9.1
☐ Closure systems to seal ducts must be installed AMI	1601.4.1	603.9.1

Max. Allowable Duct Leakage	21 IRC
☐ Rough-in test: Leakage ≤4.0 cu. ft./minute per 100 sq. ft. of conditioned floor area if air handler installed, 3.0 cu. ft./minute if not installed	1103.3.6
☐ Post-construction test: Leakage ≤4.0 cu. ft./minute per 100 sq. ft. of conditioned floor area, 8.0 cu. ft./minute if all ducts & air handlers located entirely within building thermal envelope	1103.3.6

Properly sealed ducts have a great impact on energy conservation. Legacy ducts may leak an average of 40-60% of conditioned air.

FIG. 80

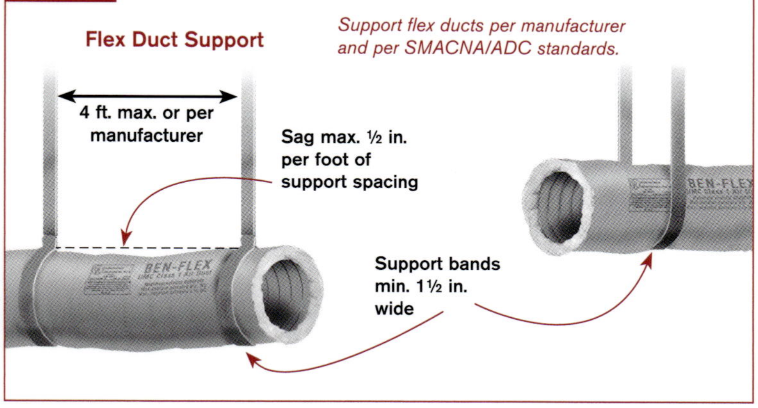

Flex Duct Installation

	21 IRC	21 UMC
☐ Install & support all ducts per SMACNA standards	1601.1.1(4)	603.3&4
☐ Flexible ducts must comply w/ UL 181 & be installed AMI	1601.4.4	603.4
☐ Ground clearance min. 4 in.	1601.4.8	603.1.3
☐ 18-in. vertical clearance where needed under duct to prevent cutting off access to crawlspace	n/a	603.2
☐ Flexible ducts in nonresidential occupancies max. 5-ft. length	n/a	603.4.1
☐ Use min. required length to make connection **F80**	1601.4.4	603.4(1)
☐ Support horizontal ducts at max. 4-ft. intervals **F80**	1601.4.4	603.4(2)
☐ Support vertical risers at max. 6-ft. intervals	1601.4.4	603.4(3)
☐ Sag max. ½ in. per foot of support spacing **F80**	1601.4.4	603.4(4)
☐ Band supports must be rigid & min. 1½ in. wide **F80**	1601.4.4	603.4(5)
☐ Duct bend radius min. 1 duct diameter centerline **F83**	1601.4.4	603.4(6)
☐ Screws not to penetrate inner liner unless AMI	1601.4.4	603.4(7)
☐ Attachment fitting collar must be beaded & min. 2 in. each side; metal worm-gear clamps must be used **F81**	1601.4.4	603.4(8)
☐ Inner liners 1 in. past bead prior to fastener **F81**	1601.4.4	603.4(9)
☐ Outer vapor barrier min. 2 wraps UL 181B tape **F81**	1601.4.4	603.4(10)

FIG. 81

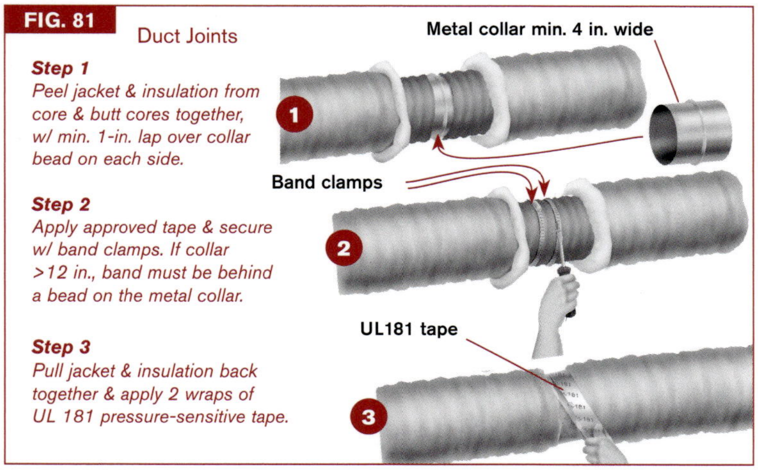

Insulation

	21 IRC	21 UMC
☐ Insulation value to be marked on factory-made ducts	1601.3	605.1
☐ Duct coverings & linings max. flame spread index 25, max. smoke-developed index 50	1601.3	605.1.2
☐ Duct coverings not to penetrate fireblocked wall or floor (UMC: Not to penetrate fire resistance–rated assembly)	1601.4.6	605.1.2
☐ Min. insulation in unconditioned space R-8 for ducts >3 in. diameter, R-6 if ≤3 in. diameter	1103.3.1	local
☐ Insulation not required in conditioned space completely within building thermal envelope & within the continuous air barrier	1103.3.2	605.1(X2)
☐ Space can be considered within thermal envelope in floor cavities, exterior walls & ceiling when air barriers & insulation separate the ducts from the unconditioned spaces	1103.3.2	local

Return Air Sources

	21 IRC	21 UMC
☐ Not from bathroom, toilet room, kitchen, mechanical room, closet, garage (IRC: Crawlspace or attic)	1602.2(4&6) & 2442.3	311.3
☐ Amount of return air from any space ≤ supply air	1602.2(2)	n/a
☐ System must be balanced by an approved method	1602.2(3)	314.1
☐ No return air from one dwelling to another dwelling unit	1602.2(7) & 2442.5	311.4
☐ Must be open to min. 25% area served	2442.3	311.3(4)
☐ Min. 10 ft. from burners or draft hoods **F82**	1602.2(1) & 2442.3	311.3(1)
☐ Return OK from room w/ fuel-burning equipment if supply air provided to replace return air, return min. 10 ft. from draft hood & room volume min. 100 cu. ft. per 1,000Btu of equipment	2442.3(5)X	311.3(6)X3
☐ Air filters required	AD106	311.2

FIG. 82

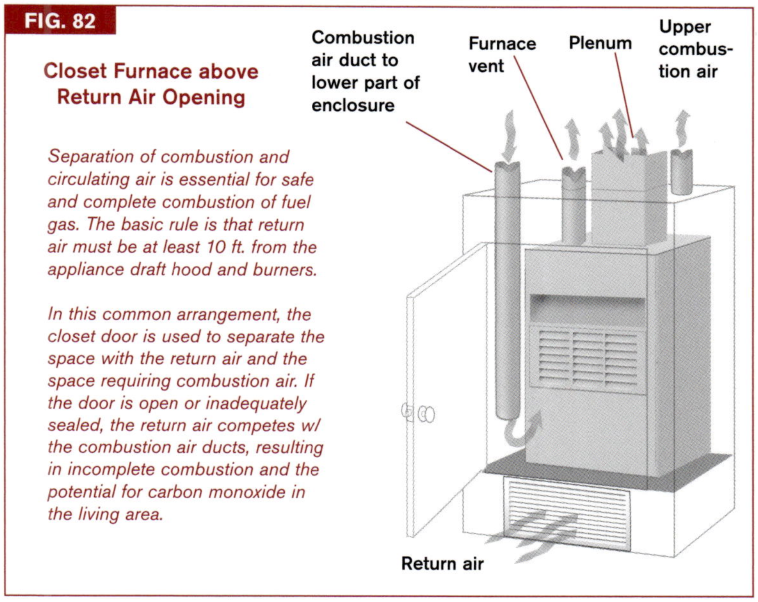

Use of Stud Cavities as Ducts

Energy codes may prohibit use of building cavities as ducts—check with local AHJ.

	21 IRC	21 UMC
☐ Stud cavities prohibited as supply air plenum	1601.1.1(7)	602.1&4.2
☐ Exposed GB in return air stud cavity limited to 50 – 125°F & moisture-controlled to prevent condensation	1601.1.1(5)	602.4.2
☐ Exposed GB requires mold/mildew-resistant surface	n/a	602.4.2
☐ Cavities must be fireblocked & sealed	1601.1.1(7)	n/a
☐ Stud cavity return not to convey from >1 floor	1601.1.1(7)	n/a
☐ Not permitted in outside walls of building envelope	1601.1.1(7)	n/a

FIG. 83

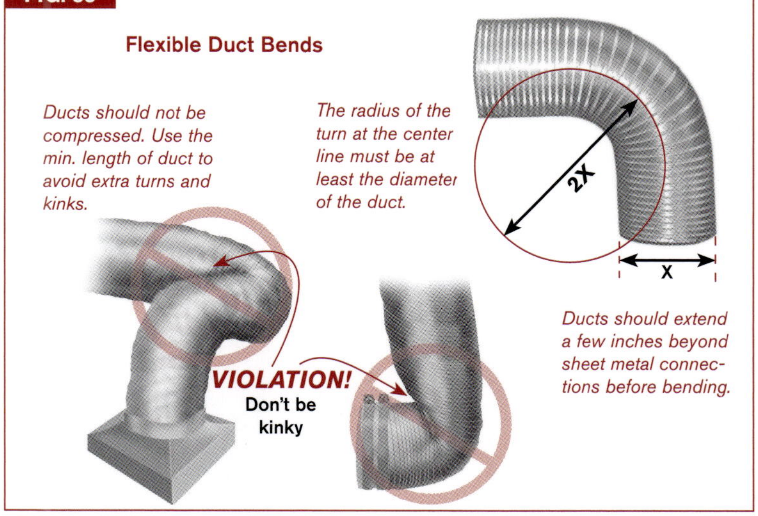

AIR CONDITIONING & HEAT PUMPS

Air-conditioning and heat-pump efficiency is measured in SEER (Seasonal Energy Efficiency Rating) or in HSPF (Heating Seasonal Performance Factor). The numbers are higher as efficiency is increased. The minimum standards for equipment sold in the United States are set by the U.S. Department of Energy and will increase slightly in 2023.

Required heating and cooling loads are determined by a calculation using ACCA Manual J. The proper sizes of equipment to meet that load are then determined by ACCA Manual S. Bigger is not better. Systems need only be sized to make up for the temperature gains or losses that occur through the building envelope; proper insulation and vapor retarders reduce the need for larger equipment. Oversized systems can lead to short-cycling, condensation, and less comfort.

Heat Pumps (HPs) & Air Conditioning (AC)	21 IRC	21 UMC
☐ Size system per ACCA Manual J & S	1103.7	1105.1
☐ Appliances must be anchored in place	1307.2	303.4
☐ Unit supported from ground on min. 3-in. raised pad **F84**	1401.4	1105.2
☐ Furnace w/ cooling coil must have blower capable of overcoming resistance of cooling coil + duct system	1411.2	904.7(1)
☐ Cooling coil downstream from heat exchangers unless L&L for upstream (e.g., SS heat exchanger)	1411.2	904.7(2&3)
☐ Central AC requires air filter	AD 106	311.2
☐ Condenser not near clothes dryer vent (California 5 ft.)	MFR	MFR
☐ Refrigerant vapor (suction) lines require insulation & vapor retarder (IRC min. R3 & max. 0.05 perm) **F84**	1411.6	1109.9
☐ Protect refrigerant piping <1½ in. from underside of roof deck from nails/fasteners (UMC: No specific distance)	1411.7	1109.7
☐ Secure refrigerant tubing within 6 ft. of condensing unit	1411.8[15]	1109.6
☐ Secure refrigerant tubing at max. 15-ft. intervals	n/a	1109.6
☐ Tamper-resistant caps outdoor refrigerant ports **F84** EXC	1411.9	1105.11
• Where protected by walls/fencing w/ key access	1411.9	1105.11X
☐ Disconnecting means within sight of condenser	4101.5	301.4
☐ GFCI-protected receptacle within 25 ft. of equipment	3901.12[16]	301.4

FIG. 84 Air-Conditioning Condenser

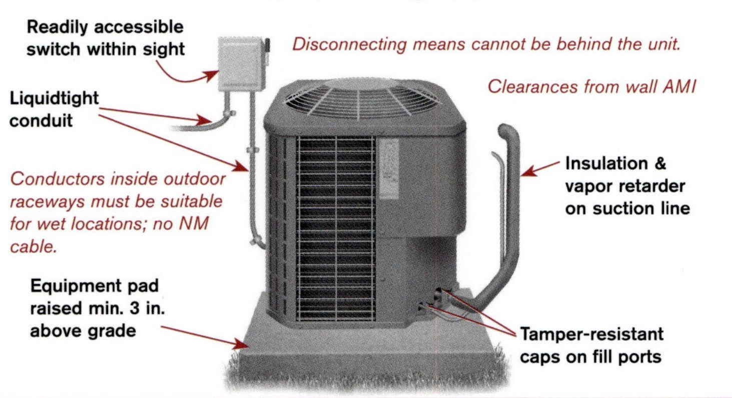

Window & Through-Wall AC Units	21 IRC	20 NEC
☐ Must have equipment grounding conductor; no adapters to existing 2-slot receptacles	3908.1	440.61
☐ Max. cord length 10 ft. if 120V, 6 ft. if 240V	MFR	440.64
☐ Cord plug OK as disconnect if controls ≤6 ft. of floor	MFR	440.63
☐ AFCI (arc-fault circuit interrupter) or LCDI (leakage current detection interrupter) or HDCI (heat-detecting circuit interrupter) required in attachment plug **F87**	MFR	440.65
☐ Max. load rating 80% of individual circuit	3702.12.1	440.62B
☐ Max. load rating 50% of shared circuit	3702.12.2	440.62C
☐ Condensate not to drain over public way	1411.3	UMC 310.1

15. Rule for securing within 6 ft. new to this IRC edition.
16. GFCI protection now required for such receptacle outlets.

EVAPORATIVE (SWAMP) COOLERS

Evaporative Coolers F85	21 IRC	21 UMC
☐ Install AMI & on level platform	1413.1(1&2)	933.4
☐ Locate so as to minimize probability of damage	1413.1	933.2
☐ Ground-mounted min. 3 in. above adjoining ground	1413.1	933.4.2
☐ Platform-mounted min. 6 in. above adjoining ground	n/a	933.4.3
☐ Provide flashing at openings into building	1413.1(3)	933.4.1
☐ OFF switch or disconnect in sight if motor >1/8 hp	4101.5	301.4
☐ Backflow protection on supply (internal air gap OK)	1413.1(4)	local
☐ Min. 10-ft. horizontal clearance to plumbing or gas vents	303.5.1	311.3(1)
☐ UMC: Electrical receptacle ≤25 ft. (IRC: Not required)	3901.12X	301.4

FIG. 85 Evaporative (Swamp) Coolers

Hot outside air is pulled through moist pads where it is cooled by evaporation and circulated through the house or building by a large blower, leaving the air cooler & slightly more humid than when it entered the cooler. These systems are suited for climates where the air is hot and humidity is low, and they are also used for makeup air in commercial kitchens.

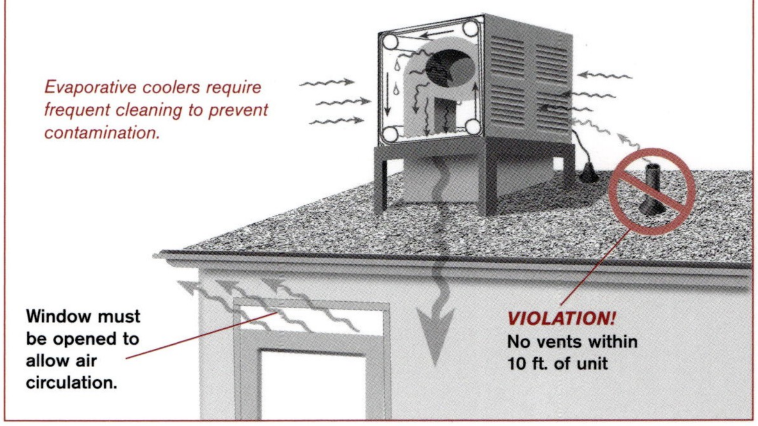

FIG. 86 Multizoned Mini-Split System

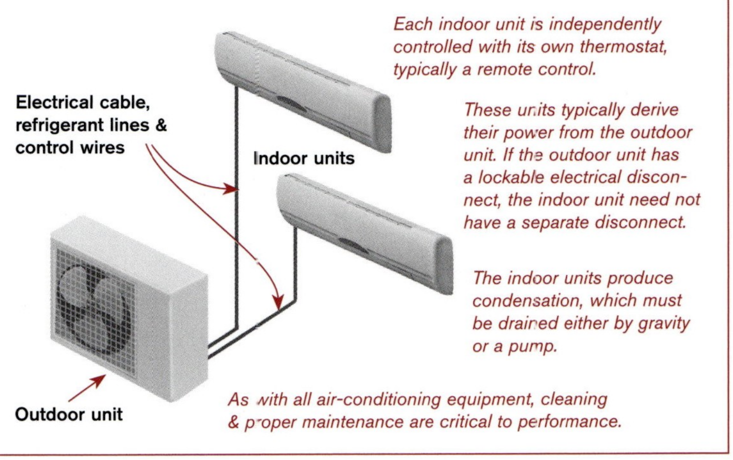

As with all air-conditioning equipment, cleaning & proper maintenance are critical to performance.

FIG. 87 Room Air-Conditioner Plug

CONDENSATE DISPOSAL

Air conditioners, Category IV high-efficiency furnaces, and tankless water heaters produce condensate, which must be disposed of to an approved location. Drains are typically connected as an indirect waste to the sanitary sewer system, though in some areas condensate is allowed to go to landscape or storm drains. Individual jurisdictions sometimes have specific directives for appliances that produce a large amount of condensate as compared to a minor amount. Condensate from high-efficiency furnaces may need to be neutralized before connecting to a sewer system. Building conditions may require use of a pump to reach an approved discharge location. Equipment located over a space where leakage would cause damage requires a secondary means to control condensate leakage.

Drain Piping

	21 IRC	21 UMC
☐ Drain pipe min. ¾ in. w/ min. ⅛-in./ft. slope **F89**	1411.3	310.1
☐ If drains from >1 unit joined together, upsize in accordance w/ approved method (UMC: 1-in. pipe if > 20 tons cooling)	1411.3.2	310.3
☐ Drain trap (in primary drain) required AMI	1401.1	310.5
☐ AC discharge to approved location (UMC: Indirectly to drywell, pit, plumbing fixture, or trapped & vented receptor)	1411.3	310.5
☐ No direct connection to waste or vent pipe	1411.3	310.1
☐ Indirect waste connection OK (lavatory tailpiece) **F90**	1411.3	310.1
☐ Not OK to drain over walkway, street, or public way	1411.3	310.1
☐ Cleanouts required to allow clearing of blockages **F89**	1411.3.3	310.3.1
☐ No threaded metal fittings into female plastic fittings	2609.2	310.7
☐ No drilling (saddle fittings) of DWV pipes to accept condensate drain (use wye tailpiece)	3003.2	UPC 310.2

Condensate Pumps

	21 IRC	21 UMC
☐ Install AMI	1411.4	310.1.1
☐ Discharge to rise vertically to gravity condensate drain	n/a	310.1.1
☐ Each appliance to have separate pump w/ interlock to prevent operation during pump failure	1411.4 & 2404.11	310.1.1
☐ Separate pumps OK to connect to single gravity indirect waste if equipped w/ check valves & approved	n/a	310.1.1

High-Efficiency Appliances (Category IV)

	21 IRC	21 UMC
☐ Provide means to collect & dispose of condensate **F88**	2427.9	802.9
☐ Condensate drain also required for Category I or III if local experience shows need (required for most tankless WH **F57**)	2427.9	802.9.1
☐ Condensate drains AMI	2427.9	802.8.3
☐ Auxiliary drain pan **F89** required if condensate stoppage could damage any building component EXC	1411.5 & 2404.10	n/a
• Automatic cutout installed in drain system	1411.5X & 2404.10X	n/a

FIG. 88 See **F102** for terminations

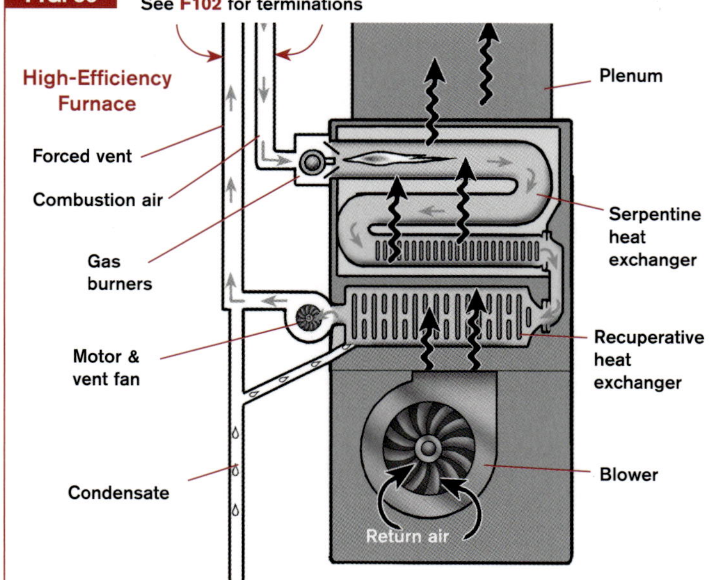

Joints in plastic vent piping require primers & must be installed AMI.

Secondary Containment F89

	21 IRC	21 UMC
☐ If condensate stoppage would damage any building components, install one of the following methods:	1411.3.1	310.2
• Water level detection in primary w/ interlocked cutout	1411.3.1(4)	310.2(1)
• Auxiliary pan w/ separate drain to readily observed location, such as over a window	1411.3.1(1)	310.2(2)
• Secondary drain piped to readily observed location	1411.3.1(2)	310.2(3)
• Auxiliary drain pan w/ leak detector & drain fitting	1411.3.1(3)	310.2(4)
☐ Equipment over drain pans above FLR of drain pan	1411.3.4	310.2.1
☐ Down-flow units w/ no secondary & no way to install auxiliary drain pan require internal blockage detector w/ interlocked cutout	1411.3.1.1	n/a

FIG. 89 Condensate Control

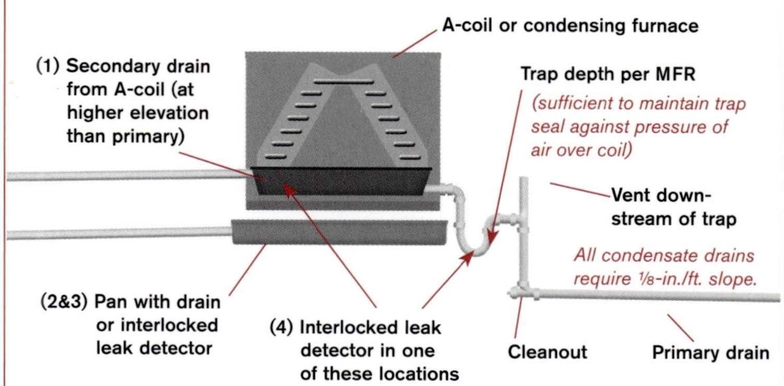

*When located over a space where blockage could damage building components, provide one of these 4 options to control secondary drainage: **(1)** pipe from secondary drain connection at A-coil, **(2)** pan with drain, **(3)** pan w/ water detection, or **(4)** blockage detection device in primary drain. The water detection device is interlocked to the controls for the unit.*

Secondary drains must discharge to a readily observable location, such as above a window.

FIG. 90

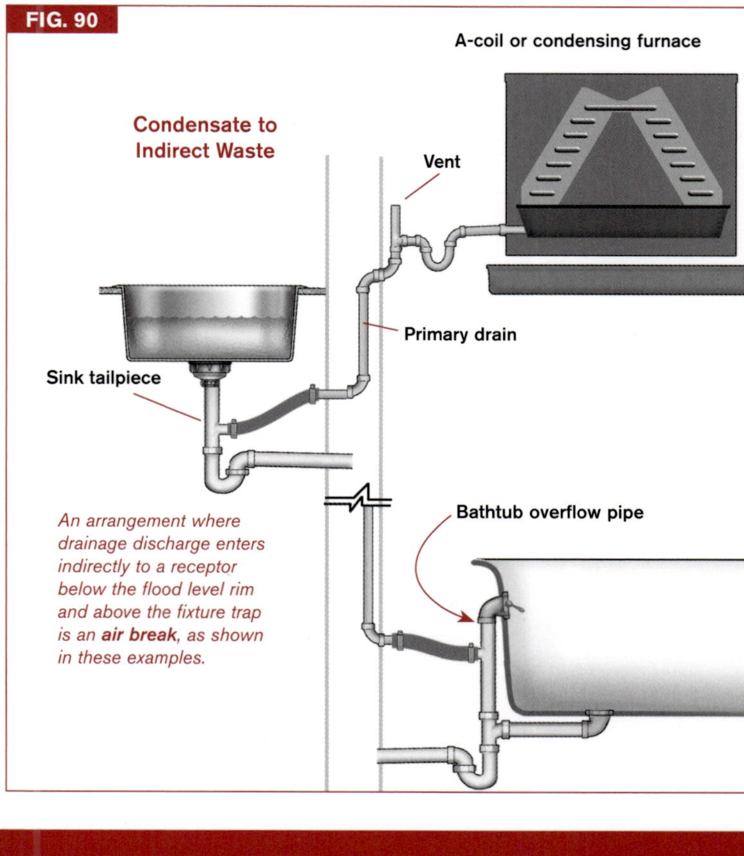

*An arrangement where drainage discharge enters indirectly to a receptor below the flood level rim and above the fixture trap is an **air break**, as shown in these examples.*

APPLIANCE ACCESS & ANCHORAGE

Appliances must remain accessible for inspection, service, repair, and replacement without the need to remove permanent construction. Listed appliances typically are marked with the required clearance from combustible surfaces.

Appliance Access

Appliance Access	21 IRC	21 UMC
☐ Maintain access to service appliances	1305.1	304.1
☐ Min. 30-in. × 30-in. level work space on control side EXC	1305.1–3	304.1
• Unit heaters & room heaters	n/a	304.1X
• Attic installations w/ access from opening **F91**	1305.1.2X1	304.4.3X
☐ Door & unobstructed passageway min. 24 in. wide to access appliances in rooms	1305.1.1	n/a
☐ Openings in attics or crawlspaces must be large enough to remove the largest appliance	1305.1.2&3	304.4

Appliance Anchorage

Appliance Anchorage	21 IRC	21 UMC
☐ Fixed appliances require anchorage in an approved manner (UMC: AMI)	1307.2	303.4
☐ WHs in SDC D 1&2FD & SDC C townhouses require strapping at upper & lower ⅓ of vertical dimension	1307.2	UPC 507.2
☐ Locate WH straps 4 in. above controls	1307.2	UPC 507.2

APPLIANCE LOCATIONS

Flood Elevation

Flood Elevation	21 IRC	21 UMC
☐ Locate above design flood elevation EXC	1301.1.1 & 1401.5	305.3
• If designed to prevent water entry & resist buoyancy	322.1.6X	305.3X
☐ If equipment replaced as part of substantial improvement, compliance w/ current standards mandatory	322.1.6	local
☐ No equipment mounted on breakaway flood walls	322.3.5(1)	305.3.1

Clearance to Combustibles

Clearance to Combustibles	21 IRC	21 UMC
☐ Clearance to combustibles AMI	1306.1, 1402.2 & 2409.3.1	303.1
☐ Clearance reduction allowed per **T52** (*p. 46*) EXC	1306.2	303.10.1
• No reductions for appliances listed for closet or alcoves	n/a	303.2
☐ Appliances not listed for closet or alcoves require room min. 12× volume of furnace, 16× volume of boiler	n/a	303.2

Appliances in Bedrooms & Bathrooms

Appliances in Bedrooms & Bathrooms	21 IRC	21 UMC
☐ Fuel-burning appliances prohibited in bedrooms, bathrooms, toilet rooms, or (IRC only: Storage closets) EXC	2406.2	904.1
• Direct-vent type installed AMI	2406.2#1	904.1(2)
• Room heaters	2406.2#2–4	916.2
• Bedroom or bathroom enclosure used for no other purpose, separated by weatherstripped self-closing door & all combustion air taken from exterior	2406.2#5	904.1(1)

Appliances in Garages

Appliances in Garages	21 IRC	21 UMC
☐ Protect appliance from impact **F56**	1307.3.1 & 2408.3X	305.1.1
☐ Ignition source min. 18 in. above floor EXC	1307.3 & 2408.2	305.1
• Flammable vapor ignition-resistant (FVIR)	1307.3X & 2408.2X	305.1
☐ Gas-fired appliance OK on garage floor if in separate space w/ access only from outside & exterior combustion air	2408.2.1	305.1.2
☐ Ducts & penetrations min. 26-gauge steel	302.5.2	local
☐ No duct openings into garage	302.5.2	local
☐ Openings around duct penetrations through house/garage separation wall sealed w/ approved materials	302.5.3	local

Rooftop Installations

Rooftop Installations	21 IRC	21 UMC
☐ Buildings > 15 ft. high must have inside means of access to roof or other means acceptable to AHJ	n/a	304.3.1
☐ Min. 30-in. × 30-in. level work space required on service side of appliance if roof slope ≥3:12 (UMC: If ≥4:12)	1305.1	304.2
☐ Guard required if < 10 ft. (< 6 ft. UMC) from roof edge	1301.1	304.3.1.1
☐ Guard min. 42 in. high	1301.1	304.3.1.1

IRC 1301.1 directs users to the IMC in the absence of a specific IRC rule. The IMC (& IBC) require a 42-in. guard for appliances less than 10 ft. from a roof edge.

Furnaces in Attics F91

Furnaces in Attics **F91**	21 IRC	21 UMC
☐ Appliance must fit through access opening	1305.1.2	304.4
☐ Opening & passageway min. 22 in. wide × 30 in. high	1305.1.2	304.4
☐ Max. 20 ft. from access opening to appliance EXC	1305.1.2	304.4.1
• If passageway ≥6 ft. high, 50-ft. OK (UMC: No limit)	1305.1.2X2	304.4.1
☐ Solid floor min. 24 in. wide to equipment	1305.1.2	304.4.2
☐ Min. 30-in. × 30-in. platform at service area EXC	1305.1.2	304.4.3
• Not if equipment can be serviced from opening	1305.1.2X1	304.4.3X
• UMC: Max. 1-ft. setback if serviced from opening	n/a	304.4.3X
☐ Floor under furnace noncombustible construction EXC	2449.4	904.3
• Not required if appliance L&L for combustible floor	2449.4	904.3X1
• Not required if floor protected in approved manner	2449.4	904.3X2

FIG. 91

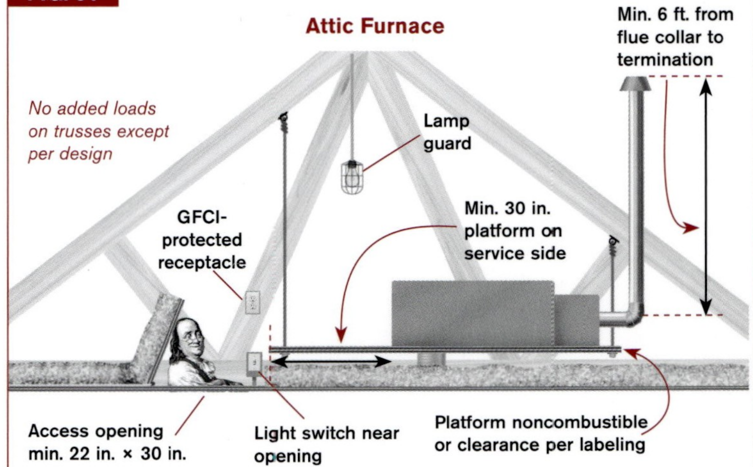

The 30-in.-deep working platform in front of the control side can be omitted if the furnace can be serviced from the opening. In the UMC, the max. setback from the opening for this exception is 12 in.

Furnaces in Crawlspaces

Furnaces in Crawlspaces	21 IRC	21 UMC
☐ Access opening min. size 22 in. × 30 in.	1305.1.3	304.4
☐ Appliance must fit through opening	1305.1.3	304.4
☐ Passageway min. 22 in. wide (UMC: Min. 24 in.)	1305.1.3	304.4.2
☐ Passageway max. 20 ft. long EXC	1305.1.3	304.4.1
• Passageway ≥6 ft. high OK for unlimited length	1305.1.3X2	304.4.1
☐ Min. 30-in. × 30-in. level space on service side **F92**	1305.1.3	304.4.3
☐ Support on concrete slab min. 3 in. above adjoining ground or suspend from floor AMI & min. 6 in. above ground **F92**	1305.1.3.1	904.3.1
☐ Excavations min. 3 in. (UMC: 6 in.) below appliance, 12 in. on sides, 30 in. on control side **F92**	1305.1.3.2	305.2
☐ If excavation >12 in. below adjacent grade, line w/ concrete extending 4 in. above adjacent grade **F92**	1305.1.3	305.2
☐ Luminaire & receptacle outlet near appliance **F92**	1305.1.3.3	304.4.4
☐ Switch for luminaire at passageway entrance	1305.1.3.3	304.4.4
☐ Protect exposed lamps by location or guards **F92**	1305.1.3.3	n/a
☐ Crawlspace lighting & receptacles require GFCI	3902.4	NEC 210.8

FIG. 92

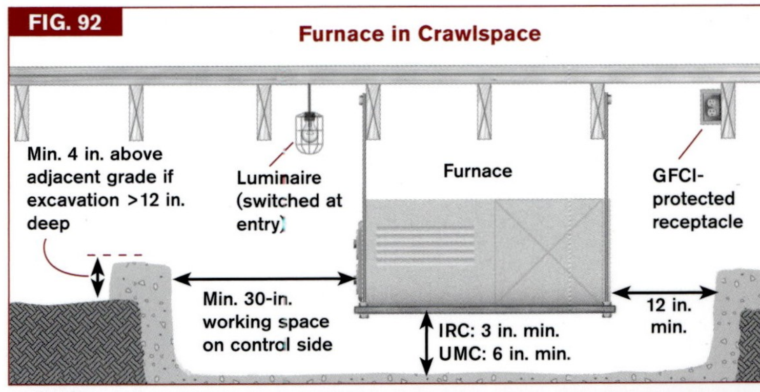

GAS APPLIANCE COMBUSTION AIR (C.A.)

As buildings have become tighter, the necessary air for combustion, ventilation, and dilution of flue gases inside a building must typically be supplied directly to the appliance space. Modern furnaces are often direct-vent and do not use the space at the appliance location as the source of combustion air. Local energy codes may have more restrictive requirements limiting the use of interior air for combustion.

General	**21 IRC**	**21 UMC**
☐ Spaces containing natural-draft appliances require C.A.	2407.1	701.1
☐ Other appliances (not Cat. I) require C.A. AMI	2407.1	701.1.1
☐ Draft hood must be in same space as appliance	2407.3	701.2
☐ Provide makeup air to offset effect of exhaust fans, such as kitchen & bath fans, dryers, etc.	2407.4	701.3
☐ Makeup air required if kitchen exhaust >400 cfm	1503.6	local
☐ Engineered installations using approved methods allowed	2407.8	701.8

Mechanically Supplied Combustion Air

☐ Mechanical C.A. supply min. 0.35 cu. ft./minute/kBtu	2407.9	701.9
☐ Appliance interlock if mechanically supplied C.A.	2407.9.2	701.9.2

Indoor Air Source

☐ Indoor air source alone only OK if infiltration ≥0.40 ACH	2407.5	701.4
☐ Min. volume of space 50 cu. ft./1kBtu/hr. **F93**	2407.5.1	701.4.1
☐ Volume includes rooms directly communicating w/ appliance space through openings w/o doors or openings per **T45,F93**	2407.5	701.4
☐ Openings connecting indoor spaces required to be located in upper & lower 12 in. of appliance space **F93**	2407.5.3.1	701.5(1)
☐ Openings connecting indoor spaces on same story min. 100 sq.in. each & min. 1 sq. in./kBtu of appliances **T93**	2407.5.3.1	701.5(1)
☐ If on different levels, min. 2 sq. in./kBtu of appliances	2407.5.3.2	701.5(2)
☐ If ACH <0.40, min. volumes for known air-infiltration method:		
• Non fan-assisted appliance (21cu. ft./ACH) per kBtu	2407.5.2	701.4.2
• Fan-assisted appliance (15 cu. ft./ACH) per kBtu	2407.5.2	701.4.2
☐ Combined indoor + outdoor air calculated by ratio of communicating spaces divided by required volume. Reduction factor of 1 minus that ratio is applied to the values in **T45**	2407.7	701.7

FIG. 93 Combustion Air from Infiltration

TABLE 45 MIN. COMBUSTION AIR OPENING SIZES

	Indoor Air[A]		Outdoor Air Openings		
Btu	Opening size[B]	cu.ft. (sq. ft.[C])	1 in./2kBtu/hr.[D]	1 in./3kBtu/hr.[E]	1 in./4kBtu/hr.[F]
30k	100 sq. in.	1,500 (188)	15 sq. in.	10 sq. in.	7.5 sq. in.
40k	100 sq. in.	2,000 (250)	20 sq. in.	13.3 sq. in.	10 sq. in.
50k	100 sq. in.	2,500 (313)	25 sq. in.	16.7 sq. in.	12.5 sq. in.
60k	100 sq. in.	3,000 (375)	30 sq. in.	20 sq. in.	15 sq. in.
80k	100 sq. in.	4,000 (500)	40 sq. in.	26.7 sq. in.	20 sq. in.
100k	100 sq. in.	5,000 (625)	50 sq. in.	33.3 sq. in.	25 sq. in.
125k	125 sq. in.	6,250 (781)	62.5 sq. in.	41.7 sq. in.	31.3 sq. in.
150k	150 sq. in.	7,500 (938)	75 sq. in.	50 sq. in.	37.5 sq. in.
200k	200 sq. in.	10,000 (1250)	100 sq. in.	67 sq. in.	50 sq. in.

A. For construction w/ known air-infiltration rate >0.40/hr.
B. Required opening between confined space (<50 cu. ft. per kBtus) & unconfined space. Refer to **F93**.
C. Example sq. ft. based on 8-ft. ceiling—use actual room volume.
D. Applies to horizontal ducts **F95**.
E. Applies to single opening method **F94**.
F. Applies to direct exterior openings (2 opening method) or to vertical ducts **F94,95,96**.

Single Permanent Opening Method F94	**21 IRC**	**21 UMC**
☐ Single direct exterior opening OK in upper 12 in. of enclosure min. 1 sq. in./3kBtu & ≥ sum of vent connectors	2407.6.2	701.6.2
☐ Appliance min. 1-in. clearance sides & back, 6-in. front	2407.6.2	701.6.2
☐ Can be ducted to exterior or spaces communicating w/ exterior (crawlspaces, attics)	2407.6.2	701.6.2

Two Permanent Openings Method F94

☐ 2 openings—locate in upper & lower 12 in.	2407.6.1	701.6.1
☐ 2 direct exterior openings min. 1 sq. in./4kBtu **T45,F94**	2407.6.1	701.6.1
☐ 2 vertical ducts min. 1 sq. in./4kBtu **T45,F95**	2407.6.1	701.6.1
☐ 2 horizontal ducts min. 1 sq. in./2kBtu **T45,F95**	2407.6.1	701.6.1

FIG. 94 Combustion Air from Direct Exterior Openings

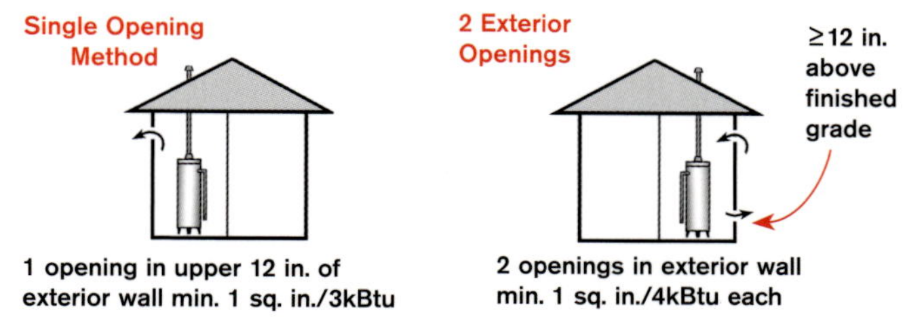

Combustion Air Ducts	**21 IRC**	**21 UMC**
☐ Zi metal or material of equivalent performance EXC	2407.11(1)	701.11(1)
• Joist/stud space OK if not >1 fireblock removed	2407.11(1)X	701.11(1)X
☐ No manual dampers or fire dampers in C.A. ducts	n/a	701.12
☐ Exterior openings ≥12 in. above finished grade **F95**	2407.11(8)	701.11(6)
☐ Duct to serve only 1 enclosure or appliance space	2407.11(3)	701.11(3)
☐ Same duct may not serve both upper & lower opening	2407.11(4)	701.11(4)

FIG. 95 Combustion Air from Ducts to Exterior

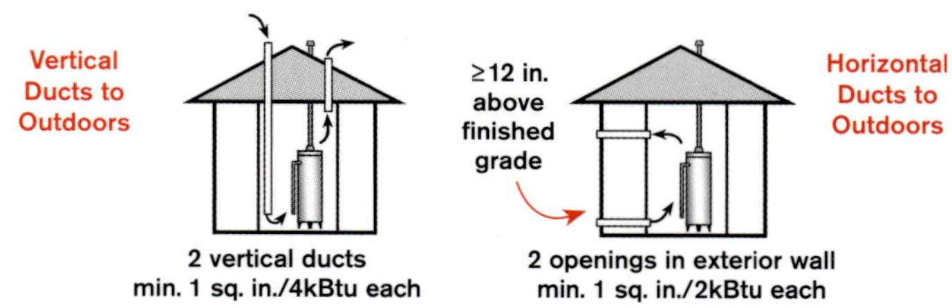

Attic & Crawlspace Sources	**21 IRC**	**21 UMC**
☐ Ventilated attics & crawlspaces considered equivalent to outdoors **F96**	2407.6.1	701.6.1
☐ No screens allowed on ducts terminating in attic	2407.11(5)	701.11(5)
☐ Crawlspace below only for lower C.A., not upper **F96**	2407.11(6)	701.11(7)

FIG. 96 Combustion Air from Openings to Spaces Equivalent to Outdoors

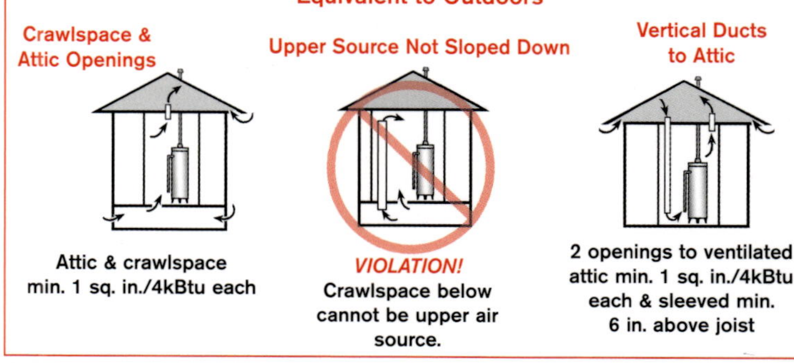

Combustion Air Openings	**21 IRC**	**21 UMC**
☐ Min. dimension of air openings 3 in.	2407.6	701.6
☐ Outside air openings require screens w/ mesh ≥¼ in.	2407.10	701.10.1
☐ Where net free area of louvers is not known, assume 75% for metal louvers, 25% for wood louvers	2407.10	701.10
☐ Nonmotorized louvers/dampers fixed in open position	2407.10	701.10
☐ Motorized louvers/dampers require appliance interlock	2407.10	701.10.2

GAS APPLIANCE CHIMNEYS & VENTS

Gas appliances at one time were almost exclusively a "gravity" vent design, where combustion gases lighter than the surrounding air rose by convection through a flue or chimney to the outdoors. As appliances became more efficient, other types of venting systems were developed, including fan-assisted appliances that induced the flow of combustion gases through a heat exchanger while still delivering these gases in a condition that is lighter than the air outside the flue. These fan-assisted appliances are still considered as "Category I" in **T46** even though they do not have draft hoods. Category IV appliances with greater efficiency extract so much heat that their vent gases are cooled to the dew point and are at positive pressure relative to the outside air. Vent systems must match the appliance category and manufacturer's recommendations.

Single-wall vents are not required to be a listed product. NFPA 54 and the UMC do not allow single-wall vents for dwellings. Single-wall vents require greater clearances and lose their heat more readily than B vents. Their flue gases may cool before reaching outdoors, resulting in a weak draft.

TABLE 46 APPLIANCE VENTING CATEGORIES ◆ NFPA 54 3.3.4.10[A]

Category	Condensation	Static Pressure	Typical Vent
I	No	Nonpositive	B Vent
II	Yes	Nonpositive	AMI
III	No	Positive	Stainless or plastic
IV	Yes	Positive	Plastic

A. See glossary for complete definitions. These definitions are also in IRC 2403 & UMC 224.0.

General

	21 IRC	21 UMC
☐ System must convey all flue & vent gases to outdoors	2427.3	802.3
☐ Choose type of vent system per **T46**	2427.4	802.4
☐ Induced-draft furnaces are "fan-assisted" Category I	2428.1	803.1
☐ B & B-W vents L&L per UL 441, Type L per UL 641	2426.1	802.1
☐ Vents must be L&L EXC	2426.1	802.1
• Single-wall metal vents & connectors	2426.1	802.7
• Plastic for Cat. IV as specified by MFR & installed AMI	2426.1	802.4.1
☐ Install all vents AMI	2426.5	802.6
☐ Appliances not required to be vented include ranges, built-in cooking units L&L for optional venting, clothes dryers (exhaust required), refrigerators, counter appliances, listed unvented room heaters	2425.8	802.2.1
☐ Aggregate total of unvented appliances max. 20 Btu/hr/cu. ft. of space containing such appliances	2425.8	802.2.2

Locations & Protection

	21 IRC	21 UMC
☐ Sheet metal shield to 2 in. above attic insulation	2426.4	802.6.1.1
☐ No screw into inner wall of double-wall vent EXC	2427.6.12	802.6(4)
• At draft-hood outlet or to single-wall connector	2427.6.12	802.6(4)
☐ Protect concealed vents <1½ in. from face of members w/ min. 16-ga steel extending min. 4 in. beyond framing inside wall	2426.7	n/a
☐ Vents not to pass through fabricated ducts or plenums	2427.3.4	802.3.5

TABLE 47 TYPE OF VENTING SYSTEM ◆ IRC T2427.4 & UMC T802.4

Appliances	Type of Vent	IRC	UMC
Listed Cat. 1, listed appliances w/ draft hoods, appliances listed for B vent	Type B gas vent	2427.6	802.6
	Chimney	2427.5	802.5
	Single-wall metal pipe[A]	2427.7	Ø 802.7.3
	Listed chimney lining for gas	2427.5.2	802.5.3
	Special vent listed for appliance (e.g., pellet stove)	2427.4.2	802.4.3
Listed vented wall furnaces	Type B-W gas vent	2427.6	802.6 907.0
Cat. II, III & IV appliances	As specified by MFR (typically plastic or SS)	2427.4.1 2427.4.2	802.4.1 802.4.3
Unlisted appliances	Chimney	2427.5	802.5
Decorative appliances in vented fireplaces	Chimney	2427.5	911.2
Direct-vent appliances	As specified by MFR	2427.2.1	802.2.6
Appliances w/ integral vent	As specified by MFR	2427.2.2	802.2.7

A. Prohibited for residential applications in UMC per section 802.7.3.

Chimneys

	21 IRC	21 UMC
☐ Inspection required before install to existing chimney	2427.5.5	802.5.7
☐ Must be lined per NFPA 211 (w/ clay or metal) EXC	2427.5.5.1[17]	802.5.7.1
• Like-for-like appliance OK to chimney passing inspection	Ø[17]	802.5.7.1X
☐ Gas appliances shall not share a chimney flue w/ a separate solid-fuel appliance	2427.5.6.1	802.5.8
☐ Vent connectors shall not be connected to a fireplace chimney flue unless fireplace flue opening permanently sealed	2427.10.13	802.10.11
☐ Chimney cross-sectional area ≥ appliance flue collar & ≤7× area of draft-hood outlet	2427.5.4(2)	802.5.5(2)
☐ Chimneys min. 3 ft. above roof & min. 2 ft. higher than part of building within 10 ft. (see *Code Check Building*)	2427.5.3	802.5.4
☐ No decorative shrouds unless L&L & AMI	2427.5.3	802.5.4.3
☐ Secure connector to prevent blocking chimney	2427.10.11	802.10.9
☐ Connector must enter chimney above extreme bottom to avoid stoppage	2427.10.11	802.10.9
☐ Connector must be secured & have thimble or stop to prevent entering so far as to restrict space to opposite wall	2427.10.11	802.10.9

TABLE 48 CLEARANCES FROM COMBUSTIBLE MATERIAL ◆ IRC T2427.10.5 UMC T802.7.3.3

Appliance	Listed Type B	Listed Type L	Single-Wall Metal Pipe
Listed appliances w/ draft hoods & appliances listed for Type B gas vents	As listed	As Listed	6 in.
Residential boilers & furnaces w/ listed gas conversion & with draft hood	6 in.	6 in.	9 in.
Residential appliances listed for Type L	NP	As listed	9 in.
Unlisted appliances w/ draft hood	NP	6 in.	9 in.
Other residential & low-heat appliances	NP	9 in.	18 in.
Medium heat appliances	NP	NP	36 in.

CATEGORY I APPLIANCE VENT SIZES

Category I appliances ship with tables for sizing the vent system. These supplied tables were developed by GAMA (the Gas Appliance Manufacturers Association). They are repeated in the model codes and downloadable from many websites. IRC Appendix AB and UMC Appendix G contain instructions and examples on how to use the tables. The tables distinguish between fan-assisted and "natural" draft appliances. Fan-assisted appliances, including those vented in common with an appliance that has a draft hood, must be sized using the tables.

Vent Size (Appliances w/ Draft Hoods)

	21 IRC	21 UMC
☐ Min. size same as flue collar	2427.6.9.1(2)	802.6.2.1(3)
☐ Max. size 7× area of smallest flue collar	2427.6.9.1(2)	802.6.2.1(3)
☐ If 2 appliances, 100% of larger + 50% of smaller	2427.6.9.1(3)	802.6.2.1(4)
☐ Offsets 45° max. except one of 60° OK	2427.6.9.2	802.6.3.2
☐ Total horizontal distance of vent + connector ≤ 75% of vertical height of vent if offsets in vent **F98**	2427.6.9.2	802.6.2.2

Vent Size Using Vent Tables

	21 IRC	21 UMC
☐ Tables can be used for all Category I appliances	2427.6.9.1(1)	802.6.2.1(1)
☐ Tables mandatory for fan-assisted Category I	2427.6.9.1(1)	802.6.2.1(2)
☐ If vertical vent diameter > than connector, vent diameter determines min. vent capacity & connector determines max. capacity	2428.2.8	803.1.8
☐ Flow area of vertical vent max. 7× appliance flue collar	2428.2.8	803.1.8
☐ Tables for single appliance to Type B double-wall only for vents not exposed to outdoors below the roof line	2428.2.9	803.1.10
☐ Type B vent in unvented chase insulated to R-8 or in unused masonry chimney flue not considered outdoors **F101**	2428.2.9	803.1.10
☐ Zero lateral values in tables only OK if straight vertical vent connects directly to a top outlet draft hood or flue collar	2428.2.4	803.1.3
☐ No elbows if using "zero lateral length" table column	2428.2.3	803.1.2
☐ Tables w/ lateral length allow for 2 90° elbows	2428.2.3	803.1.2
☐ Reduce table capacity 5% for each additional elbow up to 45° & 10% for each additional elbow >45° up to 90°	2428.2.3	803.1.2
☐ Reductions for elbows in common vents same as above	2428.3.6	803.2.5
☐ Common vent cross-sectional area ≥ largest connector	2428.3.8	803.2.7
☐ Wye/tee fitting vent connection size = common vent size	2428.3.9	803.2.9

17. The exception to the rule for "unlined chimneys that have passed inspection" was removed.

CATEGORY I APPLIANCE VENT CONNECTORS

Appliances can be directly connected to their vent, or by a vent connector. Connectors for draft hood–equipped appliances and other Category I appliances are not required to be listed. The use of single-wall connectors in cold spaces is restricted as they are likely to produce condensation under such conditions. Listed flexible vent connectors that have clearance ratings as B vents can more easily resolve difficult structural arrangements. Listed connectors that are supplied with collars to connect to B vents should not be cut or altered in the field.

General	21 IRC	21 UMC
☐ Connector required unless vent directly attached	2427.10.1	802.10
☐ Must be as straight as practical—no dips or sags	2427.10.8	802.10.6
☐ Min. ¼-in./ft. slope upwards to vent connection **F97,98**	2427.10.8	802.10.6
☐ Joints between sections & connections to draft hoods require sheet-metal screws or listed materials installed AMI	2427.10.6	802.10.5
☐ Provide adequate support	2427.10.10	802.10.8
☐ Entire connector required to have ready access	2427.10.12	802.10.10
☐ Necessary size increases required to be made only at appliance outlet connection	2427.10.3.5	802.10.2.4
☐ Connectors not to be smaller than draft-hood outlet	2428.3.17	803.2.21
☐ Connectors not >2 sizes > than flue collar diameter	2428.3.17	803.2.21

Single-Wall Connectors for Category I Appliances

☐ Not in attics or crawlspaces EXC	2427.10.2.2	802.10.1.1
• OK in other unconditioned space within exterior walls if local 99% winter design temp ≥ 5°F	2427.10.2.2X	802.10.1.1X
☐ Horizontal connector length ≤ 75% of vertical vent for draft hood–equipped appliance **F97**	2427.10.9	802.10.7.1
☐ Min. 6-in. clearance to combustibles **T48** EXC	2427.10.5	802.10.4
• Lesser clearances w/ system per **F108,T52**	2427.10.5X	802.10.4X
☐ May not pass through interior wall, floor, or ceiling	2427.10.14	802.10.12
☐ May not be covered w/ insulation	2427.10.2.3	802.10.1.2

Type B Double-Wall Vent Connectors

☐ Clearance to combustibles per L&L (1 in. typical) **T48**	2427.10.5	802.10.4
☐ Horizontal connector length ≤ 100% of vertical vent for draft hood–equipped appliance **F97**	2427.10.9	802.10.7.2

FIG. 97

Connector Length (draft-hood appliances)

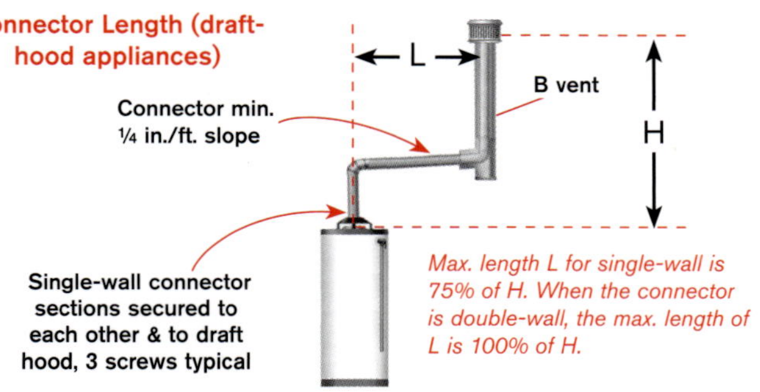

Max. length L for single-wall is 75% of H. When the connector is double-wall, the max. length of L is 100% of H.

The lengths shown in these 2 illustrations are only applicable to draft hood–equipped appliances and are not used for other Category I appliances. The vent size tables are used for those and can also be used for draft hood–equipped appliances instead of these examples.

FIG. 98

Connector to Vent with Offset (draft-hood appliances)

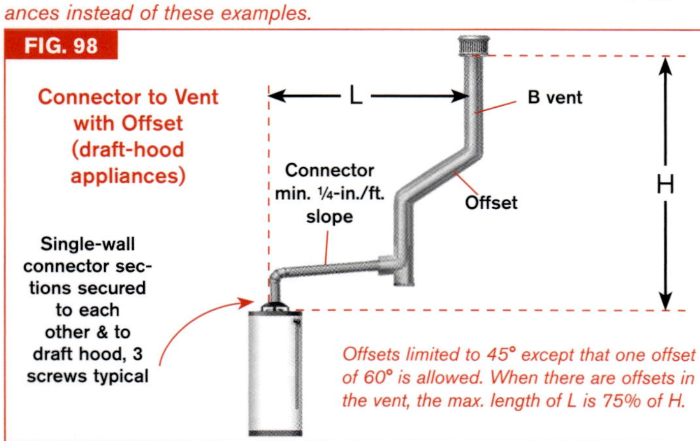

Offsets limited to 45° except that one offset of 60° is allowed. When there are offsets in the vent, the max. length of L is 75% of H.

Category I appliances may use separate connectors to a common vent. When the appliances all have draft hoods, the vent and connector sizes are allowed to be determined by the size of the draft hood outlets, or by the vent tables that are supplied with the appliance. These tables are also in the codes. If one or more appliances are fan-assisted, sizes must be determined by the tables.

Appliances Vented in Common: General	21 IRC	21 UMC
☐ Join common vent connectors as high as possible per available headroom & clearance	2427.10.3.4	802.10.2.3
☐ Two or more connectors to common vent must enter at different levels EXC **F99**	2427.10.4.1	802.10.3
• OK at same level if max. 45° from vertical **F99**	2427.10.4.1	802.10.3
☐ Smaller connector to enter above larger	2427.10.4	802.10.3.1
☐ Junctions w/ tee or wye must be MFR for purpose	2427.10.7[18]	n/a
☐ If both appliances have draft hoods, vent allowed to be sized at 100% of larger draft hood outlet + 50% of smaller & ≤7× area of smaller draft hood outlet	2427.6.9.1(3)	802.6.2.1

Common Connector Size & Length per Tables

☐ Tables mandatory for fan-assisted Category I	2427.6.9.1	802.6.2.1
☐ Route connectors using shortest possible length	2428.3.2	803.2.2
☐ Max. horizontal length of vent connector 18 in. per in. of connector diameter **T49** EXC	2428.3.2	803.2.1
• Longer lengths allowed by subtracting 10% of max. table capacity for each added multiplier of allowed length in **T49** & connector capacity based on corresponding single-appliance table as if other appliances were not present	2428.3.3	803.2.2

Example: If a 4-in. connector, normally allowed to be 6 ft., is between 6 ft. & 12 ft. in length (one multiplier of 6), reduce the allowed BTU capacity in the tables by 10%. Use the "single appliance" table for minimum capacity of each vent connector.

☐ If a single draft hood & fan-assisted appliance are vented in common, size connectors using supplied tables	2427.10.3.1	802.10.2.2
☐ Connector capacity in tables allows 2 90° elbows; subtract 5% additional elbows ≤45° & 10% for each elbow >45° & ≤90°	2428.3.7	803.2.6
☐ Max. size of vertical vent not >7× smallest flue collar	2428.3.13	803.2.17
☐ Common vent height measured from highest draft hood	2428.3.12	803.2.12

TABLE 49 MAX. HORIZONTAL LENGTH VENT CONNECTOR FOR COMMON VENTING ◆ IRC T2428.3.2 & UMC T803.2.1

Diameter (in.)	Max. Horizontal Length (ft.)	Diameter (in.)	Max. Horizontal Length (ft.)
3	4½	7	10½
4	6	8	12
5	7½	9	13½
6	9	10	15

FIG. 99

Venting in Common

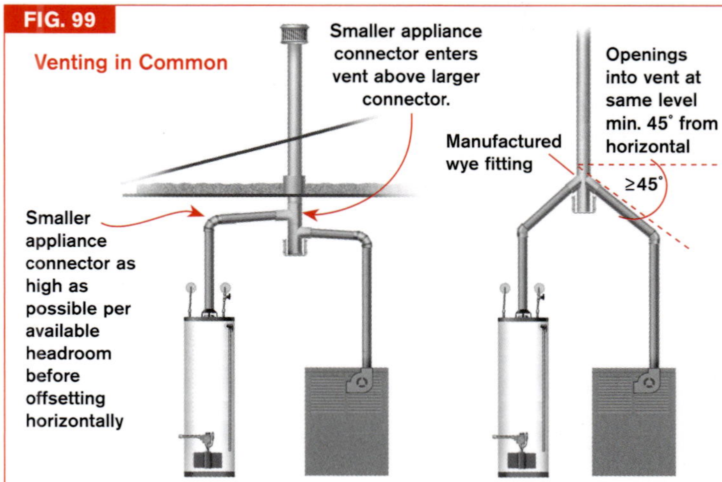

In the illustration above, assuming a B vent and single-wall connectors, IRC table 2428.3(2) or UMC table 803.2(2) or NFPA 54 table 13.2(b) would be used. Based on the Btu rating of each appliance, the tables determine the diameters of the connectors and vent for different vent heights and connector rise. The tables include allowance for up to 2 90° elbows.

18. Tees or wyes for common vents must be manufactured for the purpose, not field-fabricated.

GAS VENT TERMINATIONS

General — 21 IRC — 21 UMC

- ☐ Gas vents must extend above roof EXC ________ 2427.6.4(1&2) 802.6.2(1)
  - Direct-vent appliances **F88**,**104**,**105** __________ 2427.6.4(3) 802.6.1(1d)
  - Appliances w/ integral vents _______________ 2427.6.4(4) 802.6.2(1e)
  - Mechanical draft appliances AMI _____________ 2427.6.4(5) 802.6.2(f)
- ☐ Roof penetration requires flashing _______________ 2427.6.6 802.6.1(6)
- ☐ Must have listed cap or listed roof assembly_________ 2427.6.6 802.6.1(6)
- ☐ Decorative shrouds only if L&L & AMI _____________ 2427.6.4.1 802.6.1(5)
- ☐ Type B or L vent termination min. 5 ft. vertical above draft hood or flue collar EXC _______________ 2427.6.5 802.6.1(2)
  - Type B-W vent min. 12 ft. from bottom of wall furnace 2427.6.5 802.6.1(3)
- ☐ Min. 6 ft. when sized per vent tables _________________ 2428.2 803.0
- ☐ Vents ≤12 in. per **F100**, **T50** if > 8 ft. from wall_____ 2427.6.4(1) 802.6.2(1a)
- ☐ Vents >12 in. diameter min. 2 ft. above roof _______ 2427.6.4(2) 802.6.2(1b)
- ☐ Terminate min. 3 ft. above forced-air inlet within 10 ft. __2427.6.7 802.6.1(7)

FIG. 100

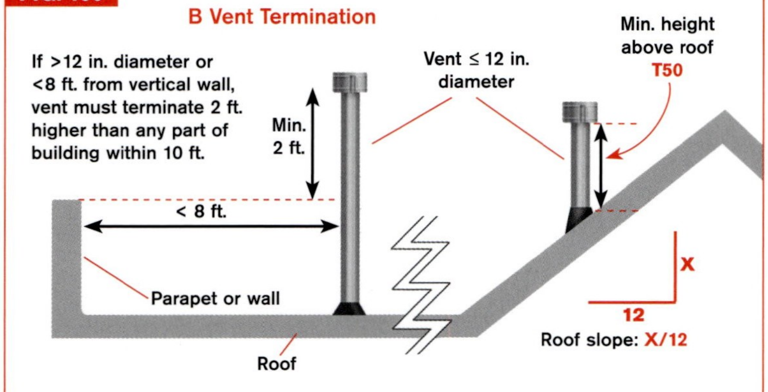

TABLE 50	B VENT TERMINATION ◆ IRC 2427.6.4 & UMC 802.6.1		
Roof Slope	Min. Height	Roof Slope	Min. Height
Flat to 6/12	1 ft.	>11/12 to 12/12	4 ft.
>6/12 to 7/12	1 ft. 3 in.	>12/12 to 14/12	5 ft.
>7/12 to 8/12	1 ft. 6 in.	>14/12 to 16/12	6 ft.
>8/12 to 9/12	2 ft.	>16/12 to 18/12	7 ft.
>9/12 to 10/12	2 ft. 6 in.	>18/12 to 20/12	7 ft. ½
>10/12 to 11/12	3 ft. 3 in.	>20/12 to 21/12	8 ft.

FIG. 101

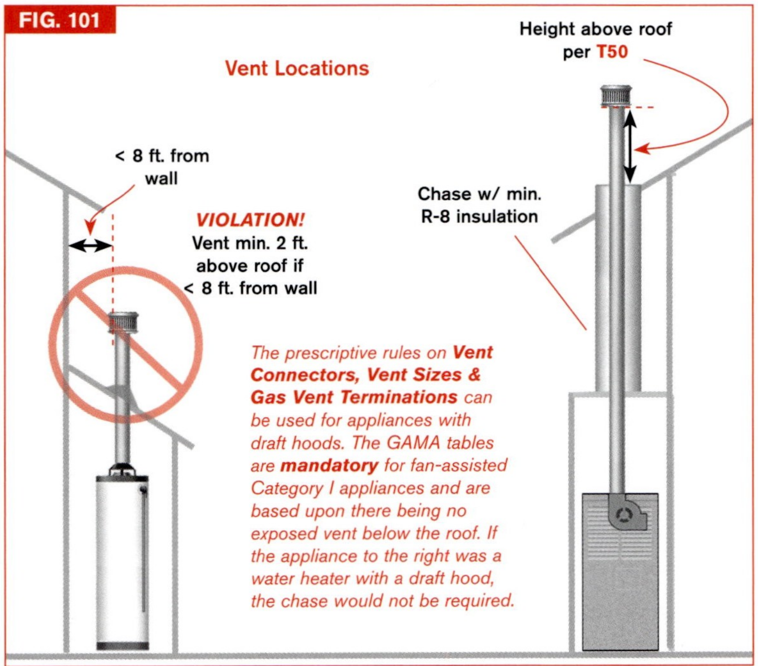

*The prescriptive rules on **Vent Connectors, Vent Sizes & Gas Vent Terminations** can be used for appliances with draft hoods. The GAMA tables are **mandatory** for fan-assisted Category I appliances and are based upon there being no exposed vent below the roof. If the appliance to the right was a water heater with a draft hood, the chase would not be required.*

Cat. IV Appliance Vents — 21 IRC — 21 UMC

- ☐ Mechanical draft systems L&L to UL 378; install AMI 2427.3.3(1) 802.3.3
- ☐ Positive-pressure systems required to be gastight__ 2427.3.3(3) 802.3.3.2
- ☐ No mixing natural & forced-draft connectors or vents 2427.3.3(4) 802.3.3.3
- ☐ Burner interlock required to forced-vent fan _______ 2427.3.3(5) 802.3.3.4
- ☐ Installation & support of vent AMI_________________ 2426.5 802.6.4
- ☐ Size Category II, III & IV appliance vents AMI _______ 2427.6.9.3 802.6.2.3
- ☐ Plastic piping requires labeling for product standard specified by appliance MFR or listing in accordance w/ UL 1738__ 2427.4.1 802.4.1
- ☐ Install plastic piping AMI per vent MFR instructions _2427.4.1.1 802.4.2
- ☐ Plastic vent joint primer must be contrasting color___ 2427.4.1.1 802.4.2

FIG. 102

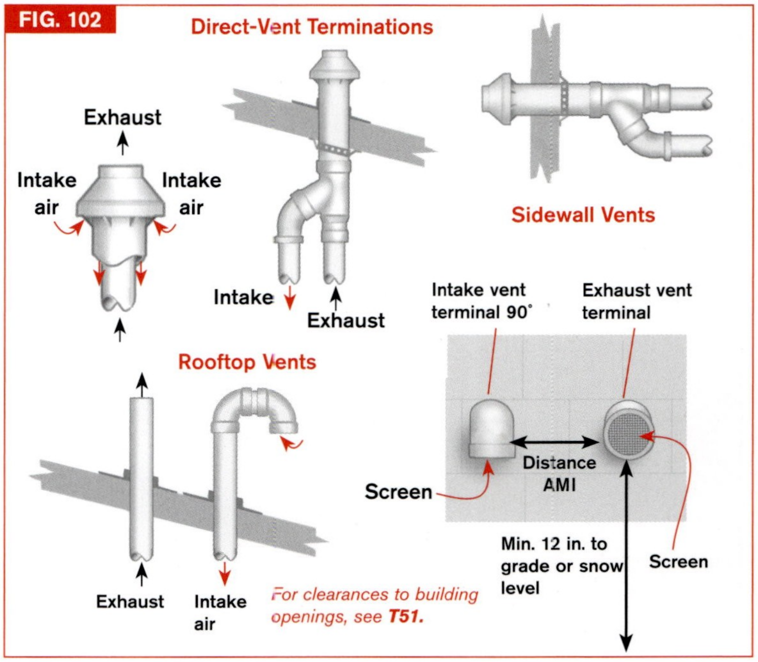

Cat. IV Vent Terminations (not Direct-Vent) — 21 IRC — 21 UMC

- ☐ Termination through wall OK **F103**,**105** clearances **T51** _ 2427.8 802.8
- ☐ Terminate min. 3 ft. above forced-air inlets within 10 ft. _2427.6.7 802.6.1(7)
- ☐ Terminate min. 4 ft. to side or below or 1 ft. above building openings, min. 1 ft. above ground level **F103**,**T51** _____________ T2427.8 802.8.1
- ☐ Through-wall vents of condensate-producing appliances not to terminate over public way or where creating nuisance___ T2427.8 802.8.3
- ☐ Min. 7 ft. above ground if adjacent to public walkway __ T2427.8 802.3.3.5
- ☐ Through-wall vent min. 10 ft. horizontally from openings in (facing) buildings if ≤ 2 ft. above or ≤ 25 ft. below openings___ T2427.8 802.8.5
- ☐ Collect & dispose of concensate from vent *(p. 39)* ____ 2427.9 802.9
- ☐ Condensate drains AMI for appliance & vent MFR _____ 2427.9 802.8.3

FIG. 103

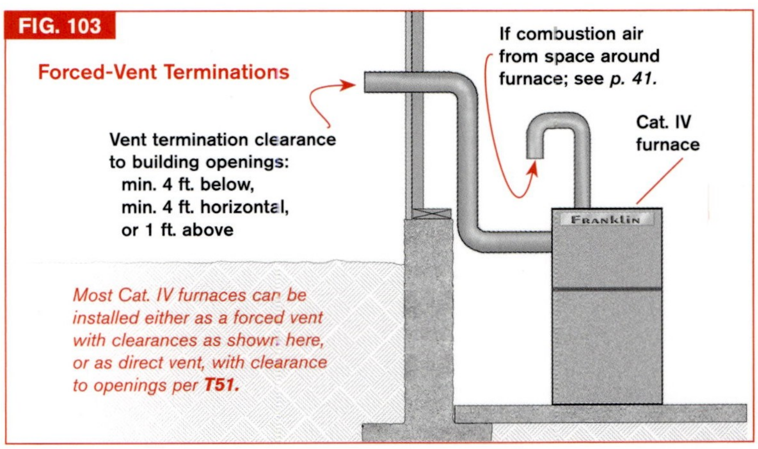

*Most Cat. IV furnaces can be installed either as a forced vent with clearances as shown here, or as direct vent, with clearance to openings per **T51**.*

DIRECT-VENT HEATING APPLIANCES

Direct-vent appliances draw their source of combustion air from the same area where they vent combustion gases. This arrangement equalizes pressure on the inlet and outlet of the firebox, which has no open flame on the building interior.

Cat. IV Vent Terminations (Direct Vent)	21 IRC	21 UMC
☐ Termination through wall OK **F102**, clearances **T51**	2427.8	802.8X1
☐ Furnaces w/ combustion air piping terminating AMI in same location as vent considered direct vent (MFR) **F102**	2427.8X	802.8X1
☐ **T51** not applicable to direct-vent combustion air intakes	2427.8X	802.8X1

Direct-Vent Gas Wall Heaters **F104**		
☐ Install listed direct-vent appliances AMI	2427.2.1	303.1
☐ Combustion air required to be from outdoors AMI	2407.1	907.1.3

FIG. 104

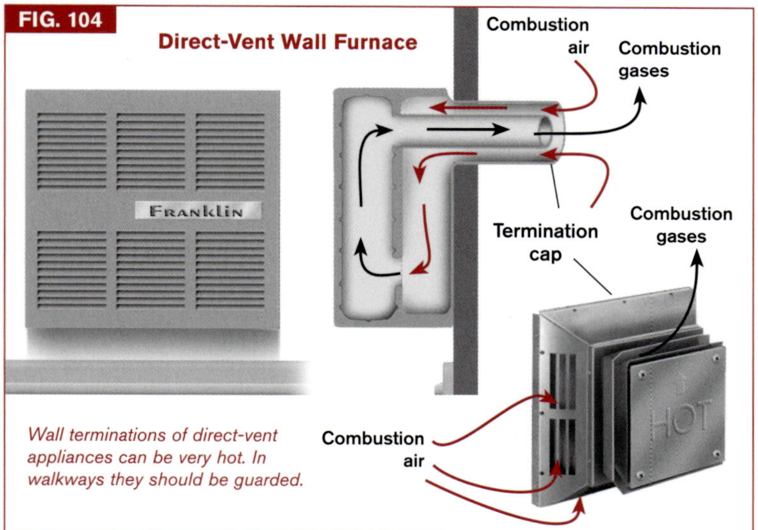

Wall terminations of direct-vent appliances can be very hot. In walkways they should be guarded.

FIG. 105

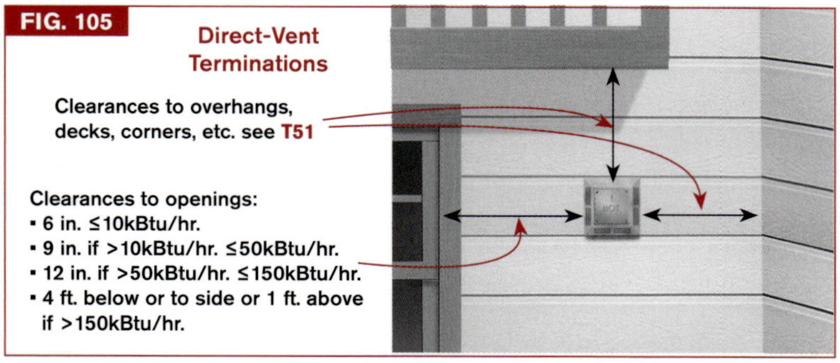

TABLE 51 — THROUGH-WALL DIRECT-VENT TERMINAL CLEARANCES
IRC T2427.8 ◆ UMC 808.8.1&2 ◆ UPC T509.8.2

Location of Vent Terminal	Min. Clearance
Clearance above finished grade, porch, deck, or balcony	12 in.
Clearance from any opening into building or to air supply inlet or combustion air inlet of appliances	6 in. for appliances ≤10kBtu/hr. 9 in. if > 10kBtu/hr. ≤50kBtu/hr. 12 in. if > 50kBtu/hr. ≤150kBtu/hr. 4 ft. below or to side or 1 ft. above if >150kBtu/hr.
Clearance to non-openable window	None
Vertical clearance to ventilated or unventilated soffit above terminal	None unless AMI
Clearance to inside or outside corner	None unless AMI
Clearance to vertical centerline of gas meter regulator vent outlet	3 ft. up to height of 15 ft. above outlet or per utility
Clearance in any direction from gas meter regulator vent outlet	3 ft. for pressures ≤2 psi, 10 ft. if >2 psi, or per utility
Clearance to mechanical air supply inlet	10 ft. horizontal or 3 ft. above
Clearance above paved sidewalk or driveway on public property	7 ft. & not located where condensate or vapor could cause nuisance or hazard
Clearance to underside of porch deck or balcony	12 in. where porch deck or balcony open on at least 2 sides. Not allowed if only one side is open.
Clearance perpendicular to wall openings in adjacent buildings	10 ft. unless 2 ft. above or 25 ft. below opening

Direct-Vent Gas Fireplaces **F106**	21 IRC	21 UMC
☐ OK in bedroom or bath if L&L & installed AMI	2406.2(1)	912.1&2
☐ Vent intake air terminal must be outdoors	2427.2.1	912.2(1&4)
☐ CSST or flex requires grommet in appliance wall	2422.1.2.3X4	1312.1
☐ Termination clearances from building openings **T51**	2427.8	802.8.2

FIG. 106

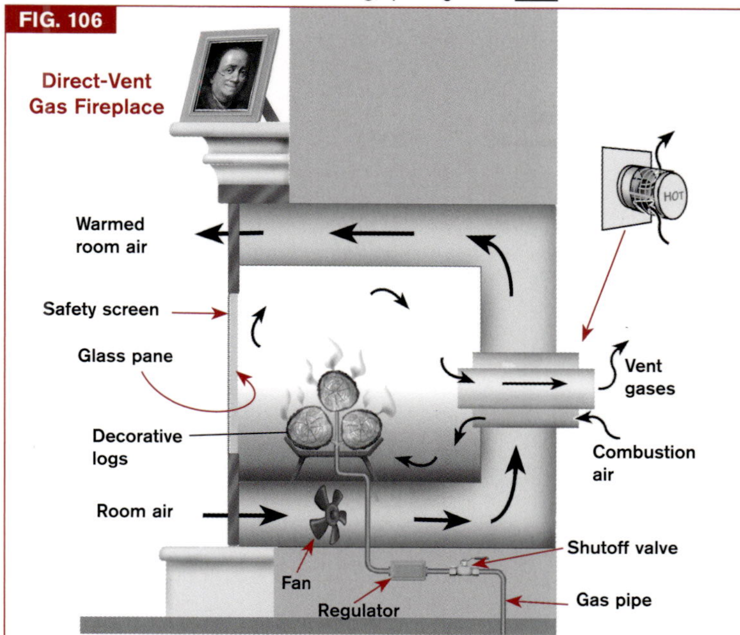

Direct-vent gas fireplaces provide heat to a room without consuming oxygen or heated air from the room. The glass front enclosure allows radiant heat to pass into the room; a fan assists ducts that pass around the firebox, and these provide warmed air to the room. The glass fronts are dangerously hot and require protective screening to avoid contact. Manufacturers offer retrofit kits for older systems that lack this safety feature.

GAS APPLIANCES IN FIREPLACES

Gas can be brought to fireplaces for log lighters to assist a solid-fuel fire, or for decorative appliances, where the primary function is the aesthetic effect of the flames. Artificial log sets inside a fireplace are decorative appliances. They do not provide heat, since the flue must be open and the heated air goes up the chimney. Direct-vent gas fireplaces, such as shown in **F106**, are now popular since they do not cause smoke pollution and they are a functional heat source.

Decorative Appliances (Log Sets) in Fireplaces	21 IRC	21 UMC
☐ Not in bath or bedroom if confined space	2432.3	911.1
☐ Maintain open vent (block damper in open position*)	MFR	911.2
☐ Thermostatic control not allowed	MFR	911.2
☐ Must be listed to ANSI Z21.60/CSA 6.26	2432.1	n/a
☐ Install AMI & per L&L	2432.1	911.2.1
☐ Fireplace screen required	MFR	911.3
☐ Appliance w/pilot or ignition system requires pilot safety	2432.2	306.1
☐ Shutoff inside firepit only if shutoff listed for same	2420.5.1	1312.6X1
☐ Shutoff in same room & within 6 ft. EXC	2420.5.1	1312.6
• Where provided w/ ready access & connector ≤6 ft.	2420.5.2	Ø
• Remote shutoffs may not serve other appliances	2420.5.2	Ø

Vented Decorative Gas Fireplaces		
☐ Not in bath or bedroom if confined space EXC	2432.3	912.1
• OK if direct vent	2406.2(1)	912.1X
☐ Must be L&L to w/ ANSI Z21.50 & be installed AMI	2434.1	912.2(1)
☐ Appliance w/pilot or ignition system requires pilot safety	2432.2	306.1
☐ Service access panels not attached to building	2434.2	912.2(3)

Vented Gas Fireplace Heaters		
☐ Comply w/ same rules as above for decorative fireplaces	2435.1	912.0
☐ Must be L&L to ANSI Z21.88 & install AMI	2435.1	912.2(1)

Log Lighters		
☐ Install log lighters AMI	2433.1	MFR
☐ Shutoff in same room & within 6 ft.	2420.5.1	1312.6
☐ No flexible connectors in firepit	2422.1.1	1312.1

Damper stops are not required for log lighters as they are assistive devices. They are only intended to start combustion of the solid fuel and then be turned off. Information on masonry fireplaces and solid-fuel burning appliances can be found in* **Code Check Building *and in* **Code Check Complete 3rd edition**.

FREESTANDING FIREPLACE STOVES

IRC section 1414.1 requires fireplace stoves to be listed, labeled and tested in accord with UL 737, which in turn references the current edition of NFPA 211. UMC section 902.10 refers to NFPA 211 for solid fuel–burning appliances. The rules for clearance, protection, and clearance reductions are virtually identical among the IRC, UMC, and NFPA 211.

Fireplace Stoves & Solid-Fuel Room Heaters — NFPA 211

- ☐ Equipment must be listed & installed per L&L or be approved by AHJ ____ 13.1
- ☐ Unlisted equipment must be approved & installed AMI ____ 13.1.1
- ☐ No unlisted equipment in mobile homes ____ 13.1.2
- ☐ Not in alcove or enclosed space <512 cu. ft. unless listed for same ____ 13.2.2
- ☐ Not OK in garages or where flammable vapors or liquids present ____ 13.2.3&4
- ☐ Listed appliances OK on combustible floors if per L&L & AMI ____ 13.5.1.1
- ☐ Noncombustible floor material 18 in. beyond stove on all sides EXC ____ 13.5.1.4
  - L&L floor protection assemblies OK AMI ____ 13.5.1.5
- ☐ Unlisted appliance floor protection 18 in. beyond on all sides, plus: ____ 13.5.1.2
  - If legs provide ≥6 in. of ventilated clearance under stove, 2-in.-thick masonry covered w/ min. 24-ga metal **F107** ____ 13.5.2.1
  - If legs provide ≥2 in. to <6 in. of ventilated clearance, min. 4-in.-thick hollow masonry + metal, cores open to allow airflow **F107** ____ 13.5.2.2
- ☐ If legs provide <2 in. clearance, floor noncombustible ____ 13.5.2.3
- ☐ Fuel storage (firewood) min. 36 in. from appliance ____ T13.6.1 note a
- ☐ 36-in. side, top & front clearance from appliance to combustibles EXC 13.6.1
  - Listed appliance clearance to combustibles AMI ____ 13.6.1.1
  - Reduced clearances OK per **T52**,**F109** ____ 13.6.2.1

Connectors

- ☐ Must be accessible for inspection, cleaning & replacement ____ 9.7.10
- ☐ Single-wall min. 18-in. clearance to combustibles **T48** EXC ____ T9.5.1.1
  - Lesser clearance w/ approved clearance reduction system **T52** ____ 9.5.1.2.1
- ☐ Not to pass through wall EXC ____ 9.7.4
  - Listed pass-through system ____ 9.7.4
  - Pass-through system constructed per NFPA 211 figure 9.7.4 ____ 9.7.4
- ☐ Maintain min. ¼-in./ft. rise from appliance collar to chimney ____ 9.7.6

FIG. 107

Fireplace Stove Clearances

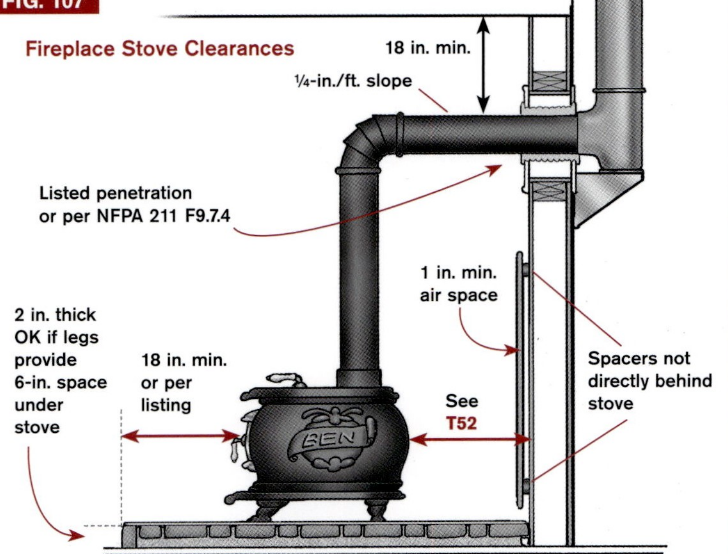

Proper clearances to combustibles are critical to the safe long-term operation of all fuel-burning systems. Improper clearances can allow heat transfer to adjacent combustible materials. Over time, heat degrades wood and lowers its ignition temperature. This process—pyrolysis—can eventually result in a fire.

Stoves & Fireplace Inserts to Masonry Fireplaces — NFPA 211

- ☐ Connector must extend to flue liner—not just to firebox ____ 13.4.5.1
- ☐ If connector enters direct through chimney wall above smoke chamber, noncombustible seal required below entry ____ 13.4.5.1
- ☐ No dilution of combustion products in flue w/ habitable space air ____ 13.4.5.1
- ☐ Flue not less than size of appliance collar ____ 13.4.5.1
- ☐ Flue diameter max. 2× appliance collar if chimney walls exposed to exterior below roof, 3× if no part exposed below roof ____ 13.4.5.1
- ☐ Installation must allow for chimney inspection & cleaning ____ 13.4.5.1

CLEARANCE REDUCTION SYSTEMS

Clearance reduction systems are used with solid fuel, oil-burning, and gas-burning appliances. While the basic rule is that appliances be installed with their required clearances to combustible materials, practical considerations must be taken into effect. The codes therefore provide a means to reduce the required clearances by the use of intermediary systems that protect the combustible construction. **T52** has the same values as T9.5.1.2 & 13.6.2.1 in NFPA 211 and NFPA 54 section 10.2.3. Though gypsum as a material is noncombustible, gypsum board is considered a combustible material. Prolonged heat exposure will cause it to decompose, and its paper backing is considered combustible.

Clearance Reduction Systems — 21 IRC

- ☐ Clearance reductions allowed per **T52** EXC ____ 1306.2, 1803.3.4, 2409.2
  - Clearance allowed based on assembly L&L to UL 1618 1306.2.1 & 2409.2
- ☐ Gas appliance & vent connector reductions per **F108**,**T52** ____ 2409.2
- ☐ Solid fuel appliances cannot be reduced to <12 in. EXC ____ 1306.2.3
  - Appliances listed for <12 in. & installed AMI ____ 1306.2.3
- ☐ No spacers directly behind appliance or connector **F107**,**109** ____ F1306.2
- ☐ Spaces noncombustible (stacked washers, conduit, etc.) ____ F1306.2
- ☐ Ventilated air space min. 1 in. & open at edges **F107–109**,**T52** ____ 1306.2
- ☐ Air space in corner open top & bottom **F109**,**T52** ____ 1306.2
- ☐ Air space on flat wall open top & bottom or side & top **F108**,**T52** ____ 1306.2

FIG. 108

Clearance Reduction for Gas Equipment or Gas Vent Connectors

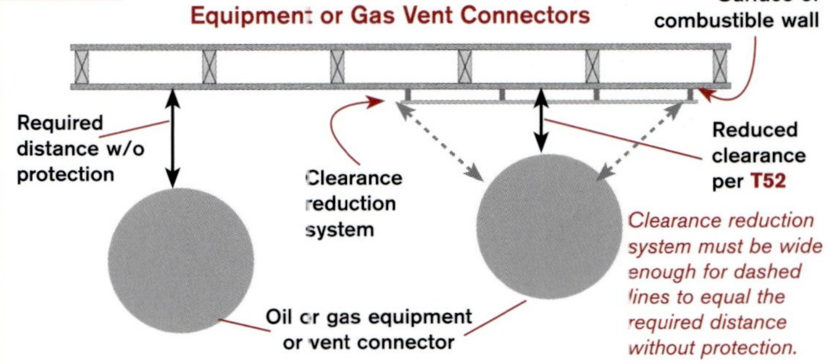

Clearance reduction system must be wide enough for dashed lines to equal the required distance without protection.

TABLE 52 — CLEARANCE REDUCTION ◆ IRC T1306.2 UMC T303.10.1

Unprotected Clearance	36	18	12	9	6	36	18	12	9	6
Protection Method	Protected Wall Clearance (in.)[A]					Protected Ceiling Clearance (in.)[A]				
3½-in. masonry w/o air space[B]	24	12	9	6	5	n/a	n/a	n/a	n/a	n/a
½-in. insulation board over 1-in. fiber or mineral wool batts[C]	18	9	6	5	3	24	12	9	6	4
24-ga Zi steel over reinforced batts w/ air space[B]	12	6	4	3	3	18	9	6	5	3
3½-in. masonry w/ air space[B]	12	6	6	6	6	n/a	n/a	n/a	n/a	n/a
24-ga Zi steel w/ air space[B]	12	6	4	3	2	18	9	6	5	3
½-in. insulation board w/ air space[B]	12	6	4	3	3	18	9	6	5	3
24-ga Zi steel w/ air space over 24-ga Zi steel w/ air space[B]	12	6	4	3	3	18	9	6	5	3
1-in. insulating batts[C] between 2 24-ga Zi steel w/ air space[B]	12	6	4	3	3	18	9	6	5	3

A. Clearances are measured in closest stretched-string distance. In no case can a solid fuel–burning appliance clearance be reduced to <12 in.

B. Air spaces must be a min. of 1 in. and be ventilated by being open at the bottom & top edges or sides & top edges. Spacers must be noncombustible & cannot be mounted directly opposite the appliance or connector.

C. Insulation fiber or mineral wool must have thermal conductivity ≤1.0 Btu/sq. ft. Mineral wool blankets or board min. density 8 lb./cu. ft. & min. melting point 1,500°F.

FIG. 109

Clearance Reduction System for Fireplace Stove

The spacers that hold the clearance reduction system out from the wall must not be located directly behind the appliance or connector. Metal fasteners can transfer heat to wood studs. The appliance clearances must comply w/ ***T52***.

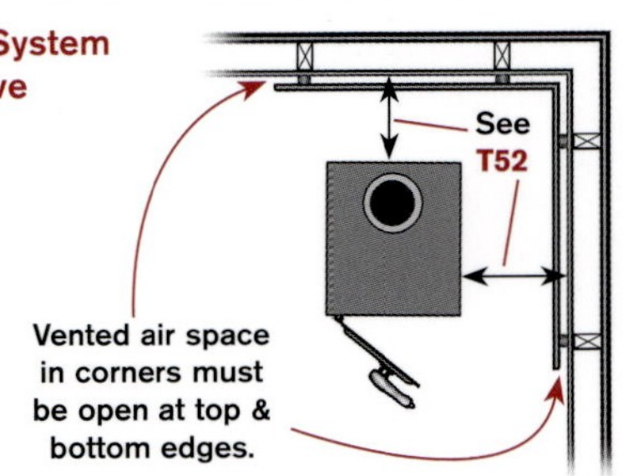

GAS WALL FURNACES

Gas-fired wall furnaces that are not direct-vent can only be located in rooms or spaces large enough to meet the combustion air requirements of the appliance. Because they use indoor air for combustion, they are typically found only in older buildings with high air-infiltration rates. Proper vent installation is very important. The clearances for type BW vent in a wall are less than for type B vent. When a wall furnace is replaced in an existing building, one side of the wall above the furnace should be opened for inspection of vent clearances and proper installation of the header plate (base plate) that connects the furnace to its vent.

Furnace Clearances	21 IRC	21 UMC
☐ From sidewall—install AMI **F110**	2436.1&3	907.2
☐ From door swing 12 in. IRC (UMC: AMI) **F110**	2436.4	907.1
☐ Do not rely on doorstops to maintain clearance	2436.4	n/a
☐ Clearance below structural projections AMI	2436.3	907.1

FIG. 110

Wall Furnace Clearances

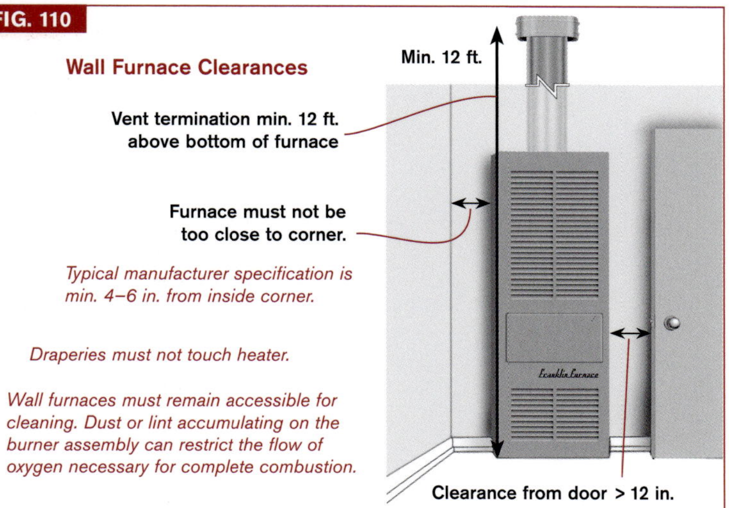

Vent Installation	21 IRC	21 UMC
☐ BW vent capacity ≥ appliance input rating	2427.6.3	802.6(2)
☐ Cut top & floor plates flush to stud **F111**	2436.2	907.1.2
☐ Ceiling plate spacers ventilate first stud space **F111**	2436.2	907.1.2
☐ Subsequent ceiling stud plates firestopped AMI **F112**	2436.2	907.1.2
☐ Single-story vent systems only single story or top floor	2436.2	907.1.2
☐ Multistory systems OK in single story or multistory	2436.2	907.1.2
☐ Approved sleeve min. 2 in. above attic insulation **F112**	2426.4	802.6.1.1
☐ Vent min. 12 ft. above bottom of furnace **F110**	2427.6.5	802.6.1(3)

FIG. 111

Wall Furnace Vent

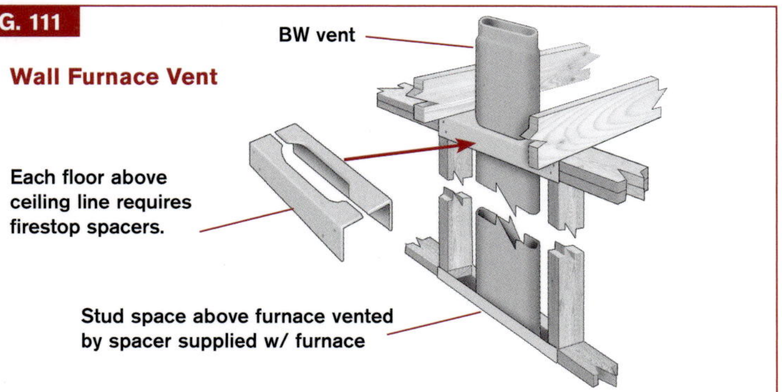

FIG. 112

BW Vent Insulation Sleeve

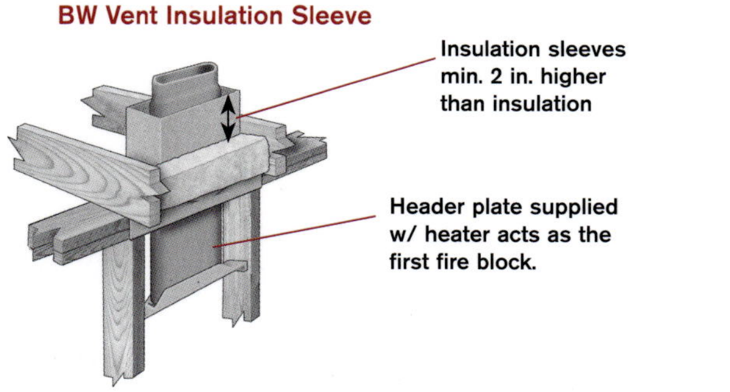

Wall Furnace Installation	21 IRC	21 UMC
☐ Must be listed (IRC: ANSI Z21.86/CSA 2.32)	2436.1	907.1
☐ Install AMI	2436.1	907.1
☐ OK in bedroom or bath if adequate combustion air	2406.2(2)	902.2
☐ Fan assist only if L&L & AMI	2436.1	907.1
☐ No ducts attached to wall furnaces	2436.5	907.1
☐ Panels, grills & access doors not attached to building	2436.6	907.1.4
☐ Header plate at top of furnace AMI	2436.1	907.1
☐ Stud bay depth AMI	2436.1	907.1
☐ Unlisted furnaces not OK in combustible construction	2404.3	907.1.1

GAS FLOOR FURNACES

See *p. 40* for rules governing underfloor access & passageways to a floor furnace. Grills for these furnaces can become dangerously hot; child-protective measures may be necessary. The furnace must be located such that a person can walk around the floor register without stepping on it directly.

Underfloor Area	21 IRC	21 UMC
☐ Must be listed (IRC: ANSI Z21.86/CSA 2.32)	2437.1	906.1(1)
☐ Install AMI	2437.1	906.1(1)
☐ Temp limit control required	2437.1	906.2(1)
☐ Unlisted furnaces allowed only in noncombustible floors & must be equipped w/ temp limit control	Ø	906.2(2)
☐ Framing around opening properly braced & headed	2437.3	906.5
Excavation Clearances		
☐ 6-in. clearance to ground (2-in. if factory sealed) **F113**	2437.4	906.7
☐ 12-in. side clearance; 18-in. on control side **F113**	2437.4	906.7
Upper Floor Installations		
☐ May not project into habitable space below	2437.5&6	906.11
☐ Projection into nonhabitable space OK if supplied w/ combustion air & enclosed w/ noncombustible materials	2437.6	906.11
☐ Projection into nonhabitable basement space OK	2437.5	906.12

FIG. 113

Floor Furnace

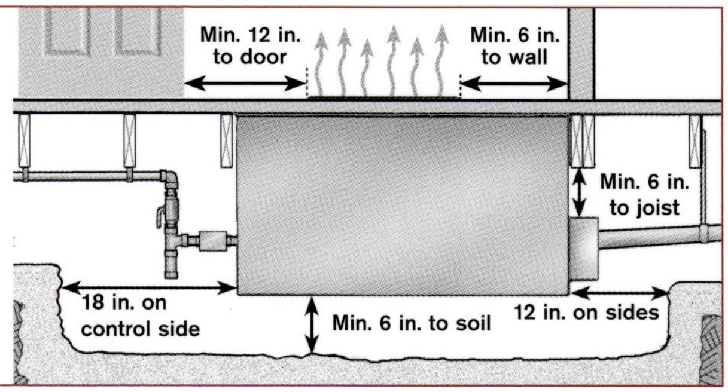

Clearances above Floor	21 IRC	21 UMC
☐ Not in doorway, landing, passageway, or exit way	2437.2(1)	906.4(1)
☐ Flat-register type min. 6 in. from wall **F115**	2437.2(2)	906.4(2)
☐ Wall-register type min. 6 in. to inner corner **F114**	2437.2(2)	906.4(2)
☐ Two adjoining sides must have 18-in. clearance	2437.2(2)	906.4(2)
☐ Min. 12 in. from door swing or draperies **F113,114**	2437.2(3)	906.4(3)
☐ Thermostat in same area as register; no intervening doors	2437.2(5)	906.1(3)

FIG. 114

Wall-Register Floor Furnace Clearances

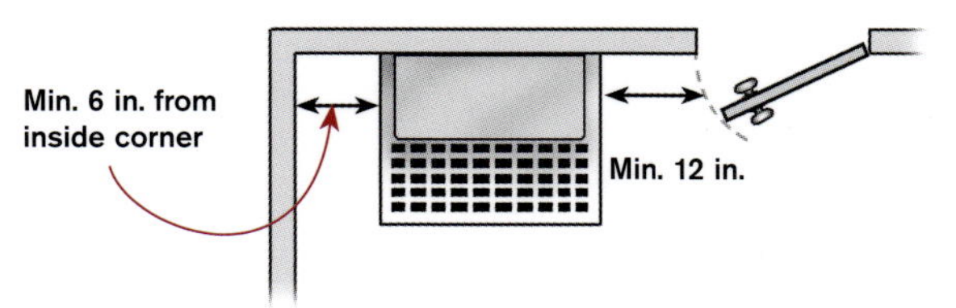

FIG. 115

Improper Floor-Register Clearances

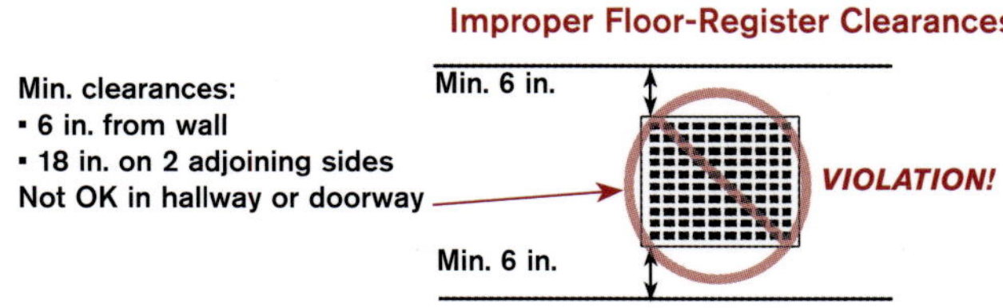

GLOSSARY

Note: The definitions below do not include the complete text that is found in the codes. Refer to IRC sections 202 and 2403, UMC and UPC Chapter 2, and the referenced standards in Table 1 for complete definitions of these and other terms.

Access (to) That which enables a device, appliance, or equipment to be reached. The necessity to first remove a door, panel, or similar obstruction is allowed, unlike equipment which has ***ready access to***, for which no obstructions are allowed.

AHJ (Authority Having Jurisdiction): An organization responsible for enforcing the code, typically the building department & its authorized representatives.

Air admittance valve (AAV): F36,37 A normally closed valve that substitutes for a vent and allows air into the drainage system under negative pressure, thereby protecting traps from siphonage.

Air break (drainage): An arrangement where discharge from a fixture, appliance, or device drains indirectly into a receptor and enters below the FLR and above the trap seal, such as a clothes washer standpipe. **F74,90**

Air conditioning: The process of heating, cooling, humidifying, dehumidifying, filtering, or otherwise treating air in a building. The most common usage of this term is to refer to cooling.

Air gap, drainage: The unobstructed vertical distance through free atmosphere between a waste pipe outlet and the FLR of the fixture or receptor into which it discharges. Air-gap devices **F67** are used for dishwasher drains, reverse osmosis filters **F68**, and water softener brine drains.

Air gap, water distribution: F42 The unobstructed vertical distance through free atmosphere between the lowest opening of water supply discharge to the FLR of a plumbing fixture.

Alcove: Aside from the common meaning of this term, it is also used to define a space such as a closet that is not "large" in relationship to the appliances within it. In the UMC, appliances in alcoves must be listed for same, and their clearances cannot be reduced. In the IRC and NFPA 54, this rule went away in the 2012 code cycle, and appliance clearances are only per the listing of the appliance. Also see "Room large in comparison to size of equipment."

Approved: Acceptable to the building official (AHJ). UL and other testing laboratories do not *approve* materials; they test products and determine their conformity to published standards. Only the BO/AHJ can approve them.

Authority Having Jurisdiction (AHJ): An organization responsible for enforcing the code, typically the building department and its authorized representatives.

Backflow: A flow of water or other liquids, mixtures, or substances into the distributing pipes of a potable supply of water from any source other than its intended source.

Backwater valve: F17-19 A device to prevent reverse flow in a drainage system.

Bathroom: In ASHRAE & the UPC, a bathroom is a room containing a bathtub, shower, or combination tub and shower. A half bath containing only a water closet & lavatory is a toilet room and is not considered a bathroom. In the NEC, a bathroom is a room containing a basin & another plumbing fixture such as a tub or toilet or shower. The IRC defines a bathroom *group* as a group of fixtures, including or excluding a bidet, consisting of a water closet, lavatory, and bathtub or shower. In a bathroom group, such fixtures are located together on the same floor level.

Branch: Any part of the piping system other than a riser, main, or stack.

Branch interval: A vertical measure of distance, 8 ft. or more, between connections of horizontal branches to a drainage stack. Measurements are taken down the stack from the highest horizontal branch connection.

Branch vent: A vent connecting 2 or more individual vents with a vent stack or stack vent. **F26**

Btu (British thermal unit): The quantity of heat necessary to raise the temperature of 1 lb. of water 1°F (roughly equal to what is given off by one stick match).

Btu/hr.: The maximum capacity of a gas-burning appliance expressed in Btu/hr.

Building drain: The lowest piping that collects discharge from all other drainage piping inside the building. It extends for 30 in. beyond the exterior walls and conveys the drainage to the building sewer. **F15**

Building official: The AHJ; an officer or other designated authority charged with administration and enforcement of the code, and its authorized representatives.

Building sewer: F15 The part of the drainage system that extends from the end of the building drain to convey its discharge to a public or private sewer or individual sewage disposal system.

Building thermal envelope: The basement walls, exterior walls, floor, roof & any other building element that encloses conditioned spaces.

Check valve: A device used to prevent the flow of liquids in a direction not intended in the design of the system. Check valves are not backflow preventers. They are often used in solar water systems and hot water circulation loops.

Chimney: A primarily vertical structure containing one or more flues for the purpose of carrying gaseous products of combustion & air from an appliance to the outside atmosphere. Factory-built chimneys must be listed & labeled. Masonry chimneys are field-constructed of solid masonry units, bricks, stones, or concrete.

Chimney connector: A pipe connecting a fuel-burning appliance to a chimney flue.

Cleanout: An access opening in the drainage system utilized for removing obstructions.

Clothes dryer, Type 1: Primarily used in family living environment. May or may not be coin-operated for public use, such as in a multifamily dwelling.

Clothes dryer, Type 2: Used in business with direct interaction with public. Usually coin-operated.

Combination waste & vent: A specially designed system of waste piping using the horizontal wet venting of one or more sinks, lavatories, or floor drains by means of a common waste and vent pipe sized to provide free movement of air above the flow line of the drain.

Combustible material: Any material not defined as noncombustible. The extent of combustibility of surface materials is measured in flame spread index & smoke-developed index. Many HVAC components require specified clearances from combustible material, including gypsum board.

Combustion air: Air required for combustion of a fuel. It includes (1) air that is burned with the fuel, (2) air for dilution of the flue gases and that is introduced into draft hoods, and (3) ventilation air that cools appliances.

Common vent (gas appliances): A vent carrying the discharge of 2 or more gas appliances to outdoors.

Common vent (plumbing): A single pipe venting two trap arms within the same branch interval, either back-to-back or one above the other. **F14,29,30**

Concealed: Not exposed to view without removal of building surfaces or finishes.

Continuous waste: A drain from two or more adjacent similar fixtures connected to a single trap. **F20**

Confined space: A room or space having a volume less than 50 cu. ft. for each 1,000 Btu input rating of all fuel-burning appliances in the room or space.

Decorative appliance for installation in fireplaces: An assembly with artificial logs & with gas burners to simulate a solid-fuel fire, installed inside a fireplace otherwise capable of burning solid fuel.

Decorative shroud: A partial enclosure for aesthetic purposes that surrounds or conceals the termination of a chimney or vent.

Direct-vent appliances: F88,F104,F106 Appliances that are constructed and installed so that combustion air and flue gases are conveyed directly from and to the outside atmosphere. These have a dual pipe system or a coaxial flue pipe inside the combustion air pipe.

Draft: The flow of gases or air through a chimney or flue caused by pressure differences. Induced draft appliances have a fan to overcome the resistance of the combustion chamber while still delivering flue gas to the vent at lower pressure relative to the atmosphere. Natural, or "gravity," draft is created by the hot flue gases being lighter than outside atmosphere. A forced-draft appliance delivers flue gas under positive pressure.

Draft hood: A nonadjustable device integral to an appliance or made part of the appliance connector. It provides for the escape of flue gases from the appliance in the event of flue blockage, allows dilution air to be introduced into the flue, prevents backdraft from entering the appliance & neutralizes the stack effect on the operation of the appliance.

Draft regulator: A device that functions to maintain a desired draft in the appliance by automatically reducing the draft to the desired value. These are usually adjustable, such as the barometric damper on an oil-burning appliance flue. A double-acting barometric draft regulator is free to move in either direction & protect against both excessive draft (that could allow the flame to lift) & backdraft.

Duct: A continuous passageway for the transmission of air (usually forced) made of factory-built components.

Energy-recovery ventilator (ERV): F77 Same as heat-recovery ventilator, with a heat-exchanger core that controls the humidity of the air being transferred.

Evaporative cooler: F85 A device used for reducing the sensible heat of air for cooling by evaporation of water into an airstream. Also known as a "swamp cooler." Evaporative coolers are used in hot, dry climates & for makeup air in commercial kitchens.

Factory-built fireplace: A fireplace composed of listed factory-built components assembled in accordance with the terms of the listing to form the completed fireplace. The appliance must be suitable for solid fuel & be equipped with a listed & properly installed chimney.

Fan-assisted appliance: An appliance equipped with an integral mechanical means to either draw or force products of combustion through the combustion chamber &/or heat exchanger.

Fireblock: Building materials installed to resist the free passage of flame to other areas through small concealed spaces of the building.

Fireplace stove: F107 A freestanding solid fuel–burning device designed to be operated with the firebox door either open or closed.

Firestop: Until the early 1990s, this term was used for what today is called fireblocking. A penetration firestop assembly is a group of materials installed to resist free passage of flame through an assembly, typically around a duct, vent, or chimney passing through a rated ceiling, floor, or wall.

Flame safeguard: A device that will automatically shut off the fuel supply to a main burner or group of burners when the means of ignition of those burners becomes disabled & when flame failure occurs.

Fixture drain: The drain from a trap of a fixture to the connection w/ any other drain pipe. Note: this is not the same as **Trap arm**.

Flood level rim: F42 The top edge of a receptor from which water overflows.

Flue: A passageway intended to carry hot gases through a chimney. The term is also used as a substitute for "vent."

Flue collar: The outlet of an appliance designed for the attachment of a draft hood, vent connector, or venting system.